Synthesis Lectures on Computer Vision

Series Editors

Gerard Medioni, University of Southern California, Los Angeles, CA, USA

Sven Dickinson, Department of Computer Science, University of Toronto, Toronto, ON, Canada

This series publishes on topics pertaining to computer vision and pattern recognition. The scope follows the purview of premier computer science conferences, and includes the science of scene reconstruction, event detection, video tracking, object recognition, 3D pose estimation, learning, indexing, motion estimation, and image restoration. As a scientific discipline, computer vision is concerned with the theory behind artificial systems that extract information from images. The image data can take many forms, such as video sequences, views from multiple cameras, or multi-dimensional data from a medical scanner. As a technological discipline, computer vision seeks to apply its theories and models for the construction of computer vision systems, such as those in self-driving cars/navigation systems, medical image analysis, and industrial robots.

Man Luo · Tejas Gokhale · Neeraj Varshney ·
Yezhou Yang · Chitta Baral

Advances in Multimodal Information Retrieval and Generation

Man Luo
AI research Scientist
Multimodal Cognitive AI Team
Intel Research Lab
Santa Clara, CA, USA

Tejas Gokhale
Department of Computer Science
and Electrical Engineering
University of Maryland, Baltimore County
Baltimore, USA

Neeraj Varshney
School of Computing, Informatics
and Decision Systems Engineering
Arizona State University
Tempe, USA

Yezhou Yang
School of Computing, Informatics
and Decision Systems Engineering
Arizona State University
Tempe, USA

Chitta Baral
School of Computing, Informatics
and Decision Systems Engineering
Arizona State University
Tempe, USA

ISSN 2153-1056 ISSN 2153-1064 (electronic)
Synthesis Lectures on Computer Vision
ISBN 978-3-031-57815-1 ISBN 978-3-031-57816-8 (eBook)
https://doi.org/10.1007/978-3-031-57816-8

© The Editor(s) (if applicable) and The Author(s), under exclusive license to Springer Nature Switzerland AG 2025

This work is subject to copyright. All rights are solely and exclusively licensed by the Publisher, whether the whole or part of the material is concerned, specifically the rights of translation, reprinting, reuse of illustrations, recitation, broadcasting, reproduction on microfilms or in any other physical way, and transmission or information storage and retrieval, electronic adaptation, computer software, or by similar or dissimilar methodology now known or hereafter developed.
The use of general descriptive names, registered names, trademarks, service marks, etc. in this publication does not imply, even in the absence of a specific statement, that such names are exempt from the relevant protective laws and regulations and therefore free for general use.
The publisher, the authors and the editors are safe to assume that the advice and information in this book are believed to be true and accurate at the date of publication. Neither the publisher nor the authors or the editors give a warranty, expressed or implied, with respect to the material contained herein or for any errors or omissions that may have been made. The publisher remains neutral with regard to jurisdictional claims in published maps and institutional affiliations.

This Springer imprint is published by the registered company Springer Nature Switzerland AG
The registered company address is: Gewerbestrasse 11, 6330 Cham, Switzerland

If disposing of this product, please recycle the paper.

Contents

1 **Introduction** 1
   1.1 Transformer-Driven Models for Language, Vision, and Multimodal Learning 3
   1.2 Multimodal Information Retrieval 4
   1.3 Multimodal Content Generation 5
   1.4 Retrieval Augmented Modeling 6
   1.5 Target Audience 7
   References 7

2 **Transformer-Driven Models for Language, Vision, and Multimodality** 11
   2.1 Vision and Language 11
      2.1.1 Communicating About What We See 12
      2.1.2 Images Invoke "Meaning" 12
   2.2 Language Modeling 13
      2.2.1 Statistical Models of Language 14
      2.2.2 Word Embeddings, Transformers, and Neural Language Models 15
      2.2.3 Transformer-Based Language Models 17
      2.2.4 Limitations of Large Language Models 21
   2.3 Modeling Techniques for Vision 21
      2.3.1 CNN 22
      2.3.2 Residual Neural Network: ResNet 23
      2.3.3 Vision Transformers (ViT) 24
   2.4 Multimodal Models 25
      2.4.1 LXMERT: Language Cross-Modal Pretraining for Vision-and-Language Understanding 27
      2.4.2 OSCAR—Object-Semantics Aligned Pretraining for Vision-and-Language Understanding 28

		2.4.3	ViLT—Vision and Language Transformer	29
		2.4.4	CLIP—Contrastive Language-Image Pretraining	30
	2.5	Broader Impact of Transformer-Driven Models for Language, Vision, and Multimodality		31
	References			32
3	**Multimodal Information Retrieval**			**35**
	3.1	Multimodal Data and Multimodal Learning		36
		3.1.1	Multimodality Fusion	37
		3.1.2	Multimodality Alignment	38
	3.2	Basic Elements of IR Systems		40
		3.2.1	Query	40
		3.2.2	Target	41
		3.2.3	Indexing	42
		3.2.4	Scoring Function	42
	3.3	Text Retrieval		43
		3.3.1	Text Representation	43
		3.3.2	Sparse Retriever	50
		3.3.3	Dense Retriever	51
		3.3.4	Hybrid Retriever	53
	3.4	Multimodal Retrieval		54
		3.4.1	Advanced Multimodal Models and Pretraining Tasks	54
		3.4.2	Mutlimodal Query for Image Retrieval	58
		3.4.3	Mutlimodal Query for Text Retrieval	58
		3.4.4	Visual and Language Dense Passage Representation	61
		3.4.5	End-to-End Multimodal Queries Retrieval	63
	3.5	Downstream Tasks Using MMIR		73
		3.5.1	VQA	76
		3.5.2	Caption Generation	77
	3.6	Evaluation		77
	3.7	Broader Impact of Multimodal Retrieval		80
	References			81
4	**Multimodal Content Generation**			**93**
	4.1	An Anthropological Lens on Visual Content Generation		94
	4.2	Conditional Image Generation		97
		4.2.1	Conditional GANs	97
		4.2.2	Score-Based Diffusion Models	98
	4.3	Taxonomy of Conditional Image Generation Tasks		99

	4.4 Categorical Conditions for Image Generation	100
	4.4.1 Category to Image	100
	4.4.2 Image Editing Using Conditional GANs	101
	4.5 Visual Conditions for Image Generation	102
	4.5.1 Semantic Labelmaps to Image	102
	4.5.2 Sketch to Image	103
	4.6 Text to Image Generation	104
	4.6.1 Text-to-Image Generation Using Generative Adversarial Networks	104
	4.6.2 Text-to-Image Generation Using Diffusion Models	105
	4.6.3 Other Variants of Text-to-Image Synthesis	108
	4.6.4 Textual Inversion	111
	4.6.5 Challenges with Evaluation of T2I Models	113
	4.7 More Applications of Text-Guided Diffusion Models	116
	4.7.1 Text to Video	116
	4.7.2 Text to Audio	118
	4.7.3 Text to 3D	120
	4.7.4 Any-to-Any Generation	121
	4.8 Image and Video Captioning	123
	4.8.1 Human-in-the-Loop Image Captioning	124
	4.8.2 Image Captioning with Commonsense Reasoning	125
	4.9 Broader Impact of Multimodal Content Generation	127
	References	128
5	**Retrieval Augmented Modeling**	**135**
	5.1 Retrieval Augmentation Architecture	136
	5.1.1 Augment the Input	136
	5.1.2 Augment the Intermediate Layers	138
	5.1.3 Augment the Output	139
	5.2 Training of Retrieval Augmented LLM	141
	5.2.1 Independent Training	141
	5.2.2 Sequential Training	142
	5.2.3 Joint Training	144
	5.3 Types of Retrieved Information	146
	5.4 Applications of Retrieval Augmented Language Models	146
	5.4.1 Open Domain Question Answering	147
	5.4.2 Use Retriever to Mitigate Hallucinations of LLMs	147
	5.4.3 Fact Checking	148
	5.4.4 Dialogue	148
	5.4.5 Slot Filling	149

	5.5	Leveraging Generation Ability of LLMs for Better Retrieval	149
		5.5.1 Retrieving Knowledge from a Language Model	149
		5.5.2 Generating Query Context to Improve Retrieval	152
	5.6	Broader Impact of Retrieval Augmented Modeling	153
	References		153
6	**Outlook**		159
	References		163

Introduction

Writing systems emerged simultaneously and independently in many ancient civilizations across the globe, including Mesopotamia, Egypt, China, India, and Mesoamerica. The invention of writing and scripts has had a tremendous impact on the trajectory of human civilization. Archaeological findings suggest the existence of much older proto-writing. Upper Paleolithic cave paintings in Europe contain dots and "Y" shaped symbols alongside paintings of animals and are conjectured to indicate the lunar mating cycle of animals—these markings are 20,000 years old and predate any other known writing or proto-writing systems [1]. The Vinca symbols (in present-day Balkans) are untranslated symbols that have been dated to be as old as the 7th millennium BCE—these symbols have been conjectured to contain information about who owned property, numerical symbols, emblems denoting communal identity or religious objects [2]. The Sumerian epic poem "Enmerkar and the Lord of Arrata" from around 1800 BCE describes a poetic version of the story of the invention of clay tablet writing systems [3]:

> Because the messenger's mouth was heavy and he couldn't repeat (the message), the Lord of Kulaba patted some clay and put words on it, like a tablet. Until then, there had been no putting words on clay.

Information storage and retrieval as concepts date back to early collections of clay tablets at Ebla (2900 BC, in contemporary Syria) and Nineveh (in present-day northern Iraq, near Mosul) which consisted of clay tablets inscribed with cuneiform writings. The Nineveh tablets, assembled by Ashurbanipal, King of the Assyrians (668–631 BC), is a collection that went beyond materials relevant to running a kingdom or a state, and the Epic of Gilgamesh was part of that collection [4]. The content of the clay tablets was multimodal in the sense that besides having graphic text (pictograms), and maps, they had physical tokens representing numbers, and also such tokens placed on clay envelops to describe the content of the envelop in a meta-data sense. The clay tablets in Ashurbanipal's collection had meta-data information recording their number in a series, filing method, and identification stamps. Multiple copies of various works in that collection suggest its simultaneous use by multiple people.

The library of Alexandria (285–145 BC) is considered perhaps the first library that used an indexing mechanism where books were sorted by alphabetical order of the first letter of the author's name. In the modern era, the Dewey Decimal Classification (DDC) debuted in 1876 and is widely used in libraries around the world. In the twentieth century, as libraries began to digitize their collections, digital library indexing systems were developed. These systems allowed for keyword searches, numeric or symbol tagging, and the use of Boolean operators. They also have made library collections accessible from any location with an internet connection. In the 21st century, there's a move towards linked data and Semantic Web technologies for libraries, enabling a higher level of interoperability and data sharing among different information systems. From ancient scroll repositories to digital databases, the history of library indexing systems shows the ever-increasing complexity of organizing and efficiently accessing information.

In the digital era, the term "information retrieval" was coined by Mooers in 1950 [5] and popularized by Fairthorne [6]. Some of the key developments in the early years of information retrieval (IR) in the digital era include development of indexing languages and their evaluations (such as the Cranfield experiments—1967), the first interactive retrieval systems such as DIALOG and MEDLINE [7], the Boolean model [8] of searching during retrieval, the formulation of "most closely match" and ranked-output [9], and the initial studies on user behavior to ground the concepts of "information need" and "relevance". The invention of the World in 1989 and DARPA's major evaluation exercises through the Text Retrieval Conferences (TRECs) that started in 1992 [10] had a huge impact on the subsequent development in the field of IR. For example, Google's early use of automated link analysis [11] to measure the relative importance of webpages and automated approaches to recognize spam, won over alternative approaches used by its competitors.

The use of machine learning methods in IR was another key landmark in the research evolution of IR. The implementation of machine learning algorithms for text classification laid the groundwork for their subsequent utilization in document ranking. Google's reverse image search, and content-based image retrieval methods in general use computer vision techniques, which now are mostly based on neural machine learning methods [12]. IR using neural methods was influenced by the vector representation of text and documents and was motivated by the need to address semantic understanding. Subsequent IR methods aimed at combining the semantic understanding and vocabulary mismatch aspect of neural IR with traditional IR challenges such as rare terms and intents [13].

Question answering (QA) is a closely related notion to IR where the query is formulated as a question in natural language, the retrieved information is answered, and the corpora (concerning which the question is answered) is often confined [14, 15]. In visual question answering (VQA), the question is asked concerning visual objects [16], and in visual-linguistic QA (VLQA) [17] and multimodal QA [18] the question is asked concerning multimodal objects. QA, IR, and prompt-based language generation [19] are getting integrated as users now expect aggregated information from multiple documents in response to their queries. This now involves the generation of text as well as images, and transformer-based large language models (LLMs) are now the key technology used for this [20, 21].

In this book, our emphasis is on multimodal information retrieval, specifically concentrating on text and image data. The traditional unimodal systems, limited to a single type of data, often fall short of capturing the complexity and richness of human communication and experience. In contrast, multimodal retrieval systems leverage the complementary nature of different data types to provide more accurate, context-aware, and user-centric search results. Text can provide specific details and context that images alone cannot convey. Conversely, images can instantly show concepts that might take longer to explain in words. Therefore, multimodal retrieval has wider applications in real world. For instance, consider you've previously visited a memorable location in New York City and captured a photo filled with landmarks and people. If you can't recall the place's name later, a multimodal system allows you to query "where is this place" along with the photo for identification. Healthcare is another domain where multimodal retrieval can be invaluable. Imagine a diagnostic support system that analyzes patients' electronic health records, which contain a mix of textual data (like doctor's notes), visual data (such as X-ray or MRI images), and even auditory data (like heart or lung sounds). A multimodal retrieval system can integrate these diverse data types to assist medical professionals in diagnosing complex conditions more accurately and swiftly.

In this book, we use the word "retrieval" in a broader sense to include the process of outputting aggregated information concerning prompted search queries to current-day generative AI models. In the rest of this chapter, we give a brief overview of the various aspects related to this. In Chap. 2, we discuss transformer-driven models for language, vision, and multimodal inputs; as transformers are key components of current-day generative AI models. In Chap. 3, we present various multimodal retrieval methods in the traditional sense of retrieval. In Chap. 4, we present generative AI models that generate multimodal content. In Chap. 5, we present how traditional retrieval can be used to augment generative models so that the resulting output is up-to-date and non-hallucinating.

1.1 Transformer-Driven Models for Language, Vision, and Multimodal Learning

This chapter focuses on the pivotal domains of artificial intelligence: language and vision, tracing their transformation through deep neural networks. The inception of convolutional neural networks (CNNs) revolutionized computer vision [22], while recurrent neural networks (RNNs) [23] marked a significant leap in natural language processing (NLP). Initially, in the early 2000s, the impact of neural networks was constrained by limited computational power and the scarcity of large-scale annotated datasets. However, this changed dramatically in 2016. Advancements in hardware enabled the training of complex models like ResNET, and the availability of datasets like ImageNet [24] showcased the true potential of deep neural networks. These advancements brought practical applications into everyday life, such as facial recognition technologies enhancing daily human interactions.

The year 2017 was a landmark in natural language processing with the introduction of the Transformer model [25]. Characterized by its self-attention, multi-head attention, and cross-attention mechanisms, the Transformer fundamentally altered the landscape of NLP. Following this, in 2018, the BERT model [26] emerged as the first language model based on the Transformer architecture, propelling NLP research to new heights. The success of BERT was underpinned by its self-supervised learning approach and deep contextual understanding of language. At its time, BERT was considered a large model, but in the context of 2023, it pales in comparison to state-of-the-art models like ChatGPT [27], which boast upwards of 175 billion parameters.

Transformers have since become the cornerstone of most cutting-edge models, not just in NLP but also in computer vision. This transition to vision was marked by the advent of the Vision Transformer (ViT) in 2020 [28]. The next frontier in AI research and application development is multimodal models. Given the multimodal nature of our world, understanding and integrating multiple forms of data is crucial for a more comprehensive understanding of our environment.

In this chapter, we will explore the evolution of both language and vision models, from their early development in the nascent years of deep learning to the contemporary era dominated by Transformer-based models. We will also discuss the most influential multimodal models that have laid the groundwork for many of the developments discussed throughout this book.

1.2 Multimodal Information Retrieval

In today's digital world, where data presents itself in myriad forms—be it text, images, videos, or a combination of these—there's an increasing need for systems that can effectively and efficiently retrieve the desired information. Multimodal Information Retrieval (MMIR) is a field that tackles the challenge of accurately retrieving specific information from a complex array of data types, including text, images, and videos, by developing systems that can efficiently search across these varied formats. In this chapter, we'll take a comprehensive journey through the landscape of MMIR, especially focusing on its applications in text-image settings.

1.2 Multimodal Information Retrieval

Initially, this chapter will outline the concepts of multimodal data and multimodal representation learning. Next, we will illuminate four key elements of Information Retrieval (IR), detailing their definitions and forms. Then we will categorize retrieval methods into two main approaches: text retrieval and multimodal retrieval. While text retrieval remains prevalent, our focus here leans more towards the dynamic field of multimodal retrieval, especially multimodal-queries retrieval, where queries seamlessly integrate components like text and images. In such systems, combining image and text information is crucial to comprehend queries and retrieve relevant documents. We will then discuss advanced multimodal transformer-based models, which transcend basic language and vision transformers. This includes a deep dive into the most exemplary models for handling multimodal queries. Following the exploration of MMIR methods, the chapter addresses their importance in key downstream applications such as question-answering and enhancing dialogue systems. Subsequent sections will investigate the evaluation metrics for IR systems, ranging from traditional metrics like precision and recall to more sophisticated measures. Finally, the chapter concludes by discussing the broader impacts of MMIR.

1.3 Multimodal Content Generation

Humans, since ancient times have observed the universe and tried to replicate it visually—in doing so, we have developed methods to create visual content. For example, cave paintings of hand prints or scenes depicting collaborative hunting tell us a story of a human community living together thousands of years ago. These images have allowed our ancestors to communicate what they saw- the environment, other creatures, other humans, and their interactions with them. Content creation, storage, and dissemination are thus an integral part of the history of our civilization.

In Chap. 4, we will learn about the research area of content generation, with special emphasis on vision-language content generation. This chapter sets up fundamental concepts in this domain such as conditional generative models and discusses several modeling techniques that use generative adversarial networks or diffusion models. We will also set up a taxonomy for conditional image generation which includes categorical conditions (e.g. using class labels as inputs to content generation models), visual conditions (such as sketches or semantic label maps), and the recent explosion of text-to-image generation (generating images directly from natural language descriptions). We will discuss text-to-image (T2I) generation in detail, by focusing on the two dominating models for T2I: GANs and diffusion models. We will learn how recent developments have resulted in many applications of T2I models in image editing, compositional generation, and iterative generation. We will also discuss several applications of text-guided generative models, for instance in generating audio, video, three-dimensional structures and assets, and other applications. We will also discuss the task of image and video captioning.

Over the last decade, sophisticated modeling strategies have emerged for image generation, language generation, audio generation, and many other forms of content generation. The models have leveraged the availability of web-scale datasets to develop training protocols that have resulted in highly realistic content generation. This is quickly creating a new wave in the digital media industry. This excitement in both academic and industrial circles has also been accompanied by challenges related to the robustness, reliability, and risks of using content-generation models. We will discuss some of the recent efforts of developing evaluation strategies and benchmarks that could potentially address some of the challenges by providing quantifiable and grounder insights about the capabilities of content generation models and their failure modes to better inform users of these technologies.

1.4 Retrieval Augmented Modeling

The previous chapters cover the motivations behind information retrieval, its fundamental principles, core components, and the various strategies employed to achieve effective retrieval. These include multimodal retrieval and generative retrieval, each with its motivations for study. Moving forward, we'll place a special emphasis on the integration of retrieval techniques with language models, a concept known as retrieval-augmented modeling.

The motivation behind studying retrieval augmented modeling is to create language models that not only understand the given input but also tap into external knowledge sources for more comprehensive and precise responses. In Sect. 5.1, we'll elaborate on the diverse ways in which language models can harness retrieved information to enhance their responses. This encompasses enriching the input to provide context, refining intermediate layers to improve comprehension, and augmenting the output for more informed responses.

The crucial aspect of retrieval-augmented modeling lies in the training of both the retriever and language models. In Sect. 5.2, we'll discuss three distinct strategies for training these models: independent training, sequential training, and joint training. In Sect. 5.3, we'll outline the various types of information that can be harnessed, such as knowledge, similar examples, and generated context, to produce informed responses. The significance here is to understand the diverse sources of information that can contribute to more accurate outputs from the model.

Shifting our focus to practical applications, Sect. 5.4 will examine the real-world impact of retrieval augmented language models. We'll explore their applications, including fact-checking and addressing the issue of factual 'hallucinations' that sometimes occur with large language models. Finally, in Sect. 5.5, we will shift our focus to leveraging the generation ability of large language models to improve retrieval performance.

1.5 Target Audience

This book covers inter-disciplinary topics, spanning information retrieval, computer vision, natural language processing, machine learning, and others. The book is intended to be a resource for advanced undergraduates, graduate students, faculty, and researchers working in these fields, adjacent areas, or those seeking an introduction to frontier research in this area. We intend to make this book accessible to readers from all of these communities to foster active dialog and exchange of ideas. Frontiers of academic research in this domain are closely connected with potential applications, such as search engines, chat-bots, AI assistants, etc. This makes the book a resource for practitioners, engineers, and designers working towards the development of such products.

In addition to this book, we have also been involved in building a community of researchers interested in adjacent topics. We were invited to organize a workshop on multimodal information retrieval at the IEEE/CVF Conference on Computer Vision and Pattern Recognition (CVPR) in June 2022. This workshop, titled "Open Domain Retrieval under Multimodal Settings" (or ODRUM for short) was designed to bring together prominent researchers from multiple research fields and perspectives such as information retrieval, natural language processing (NLP), computer vision (CV), and knowledge representation and reasoning (KRR). This workshop aimed to address the relatively nascent direction of information retrieval with queries that may come from multiple modalities (such as text, images, videos, audio, etc.), or multiple formats (paragraphs, tables, charts, etc.). The reader is encouraged to avail of the publicly available video recordings, slides, accepted papers, additional reading materials, and discussion directions that could spark open research questions in multimodal information retrieval. More details can be found at the workshop website https://asu-apg.github.io/odrum/archive_2022.html.

References

1. Bennett Bacon, Azadeh Khatiri, James Palmer, Tony Freeth, Paul Pettitt, and Robert Kentridge. An upper palaeolithic proto-writing system and phenological calendar. *Cambridge Archaeological Journal*, 33(3):371389, 2023. https://doi.org/10.1017/S0959774322000415.
2. Sarunas Milisauskas and Janusz Kruk. Middle neolithic/early copper age, continuity, diversity, and greater complexity, 5500/5000–3500 bc. *European Prehistory: A Survey*, pages 223–291, 2011.
3. Peter T Daniels. The study of writing systems. *The world's writing systems*, pages 3–17, 1996.
4. Irving Finkel. Assurbanipal's library. *Libraries before Alexandria: Ancient Near Eastern Traditions*, page 367, 2019.
5. Calvin Mooers. Information retrieval viewed as temporal signaling. In *Proceedings of the international congress of mathematicians*, volume 1, pages 572–573, 1950.
6. RA Fairthorne. Towards information retrieval. *Journal of the Operational Research Society*, 14(2):215–216, 1963.

7. R Brian Haynes, Nancy Wilczynski, K Ann McKibbon, Cynthia J Walker, and John C Sinclair. Developing optimal search strategies for detecting clinically sound studies in medline. *Journal of the American Medical Informatics Association*, 1(6):447–458, 1994.
8. Frederick Wilfrid Lancaster and Emily Gallup. Information retrieval on-line. Technical report, 1973.
9. Stephen E Robertson. The probability ranking principle in ir. *Journal of documentation*, 33(4):294–304, 1977.
10. DK Harman. Overview of the first text retrieval conference (trec-1). *NIST Special Publication*, pages 500–207, 1992.
11. Lawrence Page, Sergey Brin, Rajeev Motwani, and Terry Winograd. The pagerank citation ranking: Bring order to the web. Technical report, Technical report, stanford University, 1998.
12. Ricardo da Silva Torres and Alexandre X Falcao. Content-based image retrieval: theory and applications. *RITA*, 13(2):161–185, 2006.
13. Bhaskar Mitra, Nick Craswell, et al. An introduction to neural information retrieval. *Foundations and Trends® in Information Retrieval*, 13(1):1–126, 2018.
14. Tom Kwiatkowski, Jennimaria Palomaki, Olivia Redfield, Michael Collins, Ankur Parikh, Chris Alberti, Danielle Epstein, Illia Polosukhin, Jacob Devlin, Kenton Lee, Kristina Toutanova, Llion Jones, Matthew Kelcey, Ming-Wei Chang, Andrew M. Dai, Jakob Uszkoreit, Quoc Le, and Slav Petrov. Natural questions: A benchmark for question answering research. *Transactions of the Association for Computational Linguistics*, 7:452–466, 201. https://doi.org/10.1162/tacl_a_00276. URL https://aclanthology.org/Q19-1026.
15. Danqi Chen, Adam Fisch, Jason Weston, and Antoine Bordes. Reading Wikipedia to answer open-domain questions. In *Proceedings of the 55th Annual Meeting of the Association for Computational Linguistics (Volume 1: Long Papers)*, pages 1870–1879, Vancouver, Canada, 2017a. Association for Computational Lingu. https://doi.org/10.18653/v1/P17-1171. URL https://aclanthology.org/P17-1171.
16. Stanislaw Antol, Aishwarya Agrawal, Jiasen Lu, Margaret Mitchell, Dhruv Batra, C. Lawrence Zitnick, and Devi Parikh. VQA: visual question answering. In *2015 IEEE International Conference on Computer Vision, ICCV 2015, Santiago, Chile, December 7-13, 2015*, pages 2425–2433. IEEE Computer Society, 2015. https://doi.org/10.1109/ICCV.2015.279. URL https://doi.org/10.1109/ICCV.2015.279.
17. Shailaja Keyur Sampat, Yezhou Yang, and Chitta Baral. Visuo-linguistic question answering (vlqa) challenge. In *Findings of the Association for Computational Linguistics: EMNLP 2020*, pages 4606–4616, 2020.
18. Yingshan Chang, Mridu Narang, Hisami Suzuki, Guihong Cao, Jianfeng Gao, and Yonatan Bisk. Webqa: Multihop and multimodal qa. In *Proceedings of the IEEE/CVF Conference on Computer Vision and Pattern Recognition*, pages 16495–16504, 2022a.
19. Pengfei Liu, Weizhe Yuan, Jinlan Fu, Zhengbao Jiang, Hiroaki Hayashi, and Graham Neubig. Pre-train, prompt, and predict: A systematic survey of prompting methods in natural language processing. *ACM Computing Surveys*, 55(9):1–35, 2023.
20. Ashish Vaswani, Noam Shazeer, Niki Parmar, Jakob Uszkoreit, Llion Jones, Aidan N. Gomez, Lukasz Kaiser, and Illia Polosukhin. Attention is all you need. In Isabelle Guyon, Ulrike von Luxburg, Samy Bengio, Hanna M. Wallach, Rob Fergus, S. V. N. Vishwanathan, and Roman Garnett, editors, *Advances in Neural Information Processing Systems 30: Annual Conference on Neural Information Processing Systems 2017, December 4-9, 2017, Long Beach, CA, USA*, pages 5998–6008, 2017a. URL https://proceedings.neurips.cc/paper/2017/hash/3f5ee243547dee91fbd053c1c4a845aa-Abstract.html.
21. Tom B. Brown, Benjamin Mann, Nick Ryder, Melanie Subbiah, Jared Kaplan, Prafulla Dhariwal, Arvind Neelakantan, Pranav Shyam, Girish Sastry, Amanda Askell, Sandhini Agarwal, Ariel Herbert-Voss, Gretchen Krueger, Tom Henighan, Rewon Child, Aditya Ramesh,

References

Daniel M. Ziegler, Jeffrey Wu, Clemens Winter, Christopher Hesse, Mark Chen, Eric Sigler, Mateusz Litwin, Scott Gray, Benjamin Chess, Jack Clark, Christopher Berner, Sam McCandlish, Alec Radford, Ilya Sutskever, and Dario Amodei. Language models are few-shot learners. In Hugo Larochelle, Marc'Aurelio Ranzato, Raia Hadsell, Maria-Florina Balcan, and Hsuan-Tien Lin, editors, *Advances in Neural Information Processing Systems 33: Annual Conference on Neural Information Processing Systems 2020, NeurIPS 2020, December 6-12, 2020, virtual*, 2020a. URL https://proceedings.neurips.cc/paper/2020/hash/1457c0d6bfcb4967418bfb8ac142f64a-Abstract.html.

22. Alex Krizhevsky, Ilya Sutskever, and Geoffrey E. Hinton. Imagenet classification with deep convolutional neural networks. In Peter L. Bartlett, Fernando C. N. Pereira, Christopher J. C. Burges, Léon Bottou, and Kilian Q. Weinberger, editors, *Advances in Neural Information Processing Systems 25: 26th Annual Conference on Neural Information Processing Systems 2012. Proceedings of a meeting held December 3-6, 2012, Lake Tahoe, Nevada, United States*, pages 1106–1114, 2012. URL https://proceedings.neurips.cc/paper/2012/hash/c399862d3b9d6b76c8436e924a68c45b-Abstract.html.

23. Sepp Hochreiter and Jürgen Schmidhuber. Long short-term memory. *Neural computation*, 9(8):1735–1780, 1997.

24. Jia Deng, Wei Dong, Richard Socher, Li-Jia Li, Kai Li, and Fei-Fei Li. Imagenet: A large-scale hierarchical image database. In *2009 IEEE Computer Society Conference on Computer Vision and Pattern Recognition (CVPR 2009), 20-25 June 2009, Miami, Florida, USA*, pages 248–255. IEEE Computer Society, 2009. https://doi.org/10.1109/CVPR.2009.5206848

25. Ashish Vaswani, Noam Shazeer, Niki Parmar, Jakob Uszkoreit, Llion Jones, Aidan N. Gomez, Lukasz Kaiser, and Illia Polosukhin. Attention is all you need. In Isabelle Guyon, Ulrike von Luxburg, Samy Bengio, Hanna M. Wallach, Rob Fergus, S. V. N. Vishwanathan, and Roman Garnett, editors, *Advances in Neural Information Processing Systems 30: Annual Conference on Neural Information Processing Systems 2017, December 4-9, 2017, Long Beach, CA, USA*, pages 5998–6008, 2017b. URL https://proceedings.neurips.cc/paper/2017/hash/3f5ee243547dee91fbd053c1c4a845aa-Abstract.html.

26. Jacob Devlin, Ming-Wei Chang, Kenton Lee, and Kristina Toutanova. BERT: Pre-training of deep bidirectional transformers for language understanding. In *Proceedings of the 2019 Conference of the North American Chapter of the Association for Computational Linguistics: Human Language Technologies, Volume 1 (Long and Short Papers)*, pages 4171–4186, Minneapolis, Minnesota, 2019a. Association for Computational Linguistics. https://doi.org/10.18653/v1/N19-1423. URL https://aclanthology.org/N19-1423.

27. Partha Pratim Ray. Chatgpt: A comprehensive review on background, applications, key challenges, bias, ethics, limitations and future scope. *Internet of Things and Cyber-Physical Systems*, 2023.

28. Alexey Dosovitskiy, Lucas Beyer, Alexander Kolesnikov, Dirk Weissenborn, Xiaohua Zhai, Thomas Unterthiner, Mostafa Dehghani, Matthias Minderer, Georg Heigold, Sylvain Gelly, et al. An image is worth 16x16 words: Transformers for image recognition at scale. arXiv preprint arXiv:2010.11929, 2020.

Transformer-Driven Models for Language, Vision, and Multimodality

2.1 Vision and Language

One of the main functions of natural language is communication. Humans use language as a means of communicating their observations of the real world, their understanding of these observations, and their opinions about it. Expressing what we see in language thus offers us a way to convert visual signals into a symbolic language—with syntax, grammar, structure, and semantics that are shared and understood by communities. Language thus allows us to narrate our visual observations to others. Vision and language are inherently connected.

Such knowledge transmission has taken place in various verbal forms such as tales, plays, songs, drama, and theatre. But oral traditions require an unbroken chain of humans for transmission. Writing as a tool removed that requirement. Knowledge could now be stored forever—in symbols. The invention of printing enabled an automation of the process of writing—this led to a massive explosion of written material in China in 650 CE and later in Europe in the second millennium. Just like the digitization of images, written manuscripts, books, letters, journals, and encyclopedias have also been rapidly digitized and stored on the internet.

Today, the pace of digitization is at an all-time high—photographs and text are primarily stored in digital form with a large proportion of it on the internet. This process has resulted in a rich and diverse collection of human-authored text and image data (often paired data, i.e. text describing each image) to be collected assembled, and shared. Recently such data has been successfully used to train machine learning models that can process and understand vision and language and perform complex tasks like automated image captioning, automated question answering about images, predicting future events in videos, and retrieving images and text given a search query. This era of applications powered by models jointly trained with image and text data has ushered in a new era in technology.

In this chapter, we will learn about the modeling and learning techniques that drive multimodal applications. We will focus specifically on the recent advances in transformer-based modeling for natural language understanding, and image understanding, and how these approaches connect for jointly understanding combinations of language and image.

2.1.1 Communicating About What We See

One of the central questions in computer vision is that of "understanding" images. But what exactly do we mean by "understanding"? Different sub-areas within computer vision answer this question by taking different approaches. Let us try to understand these approaches with an example.

2.1.2 Images Invoke "Meaning"

Consider the image in Fig. 2.1. What do you observe? What thoughts does this image invoke? Let's list down several observations.

To begin with, most of us would have observed that there are two humans in this image and both of them are sitting in boats or kayaks in a water body. Each of them is holding a paddle. The human on the left is looking at the camera and smiling. The human on the right seems to be adjusting the paddle. Meanwhile, there are also several objects in the background—trees,

Fig. 2.1 What thoughts does this image invoke? *Image credits* Tejas Gokhale

bridges, buildings, a fountain, and also a TV tower. On close observation, we can also see other tiny boats in far in the background. The sky has clouds, the water has waves.

At a very fundamental level, one could say that there are many colors—the water and the sky are blue, the trees are green, the boats are yellow, and the life jackets are red. Each of these objects also has a distinct shape, size, and pose (or orientation concerning the camera). There are also optical properties such as reflection, shadows, etc. that we can observe. We can also count the number of objects—two boats, two humans, two paddles, two bridges, one fountain, and so on. In some cases, we could also read the characters printed on the boat.

Beyond these visually obvious observations, humans also tend to interpret and speculate about this image. For instance, one may reason about the emotions of the humans—the fact that the person on the left is smiling implies happiness. We could also reason about capabilities—perhaps a seasoned paddler may look at the person on the right and have doubts about their kayaking abilities.

Meanwhile, some of us may recognize the scene as being from a particular location—in this case, Pittsburgh. Using our lived experience and knowledge, we may be able to say that the image was captured in Downtown Pittsburgh on the Ohio River. This may help us answer questions about the image—for instance, where could the humans have acquired the kayaks? We could speculate whether or not the humans live in the buildings seen behind. Thus based on the level of knowledge of the viewer, the image can have different interpretations and those interpretations can help us understand the image better. A friend of the authors recognized the image as "the place where I met my wife for the first time"—this is a very personal connection to a scene that only humans can have. As the proverb goes—*"beauty lies in the eye of the beholder"*. In the case of visual understanding, *"meaning lies in the eye of the beholder"*.

2.2 Language Modeling

There is a long history of language modeling from theories of linguistics. The *structuralist* theory of linguistics focused on the systematic structure of natural language, such as developing techniques to analyze rules, patterns, grammar, and phonology of several human languages. The *generative linguistics* paradigm was championed by linguists such as Noam Chomsky with the claim that all languages shared the same common structure—a *"universal grammar"*. This theory was also influential in the development of the theory of "formal languages" which has had a huge impact on computer science. Other theories include *functionalism*—which focuses on social and communicative functions of language, especially how language is used as a tool for communicating meaning under various contexts; and *cognitive linguistics*—which connects language to cognitive psychology and holds that language emerged from cognitive faculties.

So how is natural language modeled computationally? In this chapter, we will learn about various proposals for language modeling.

2.2.1 Statistical Models of Language

Word n-gram language models [1] are statistical models of natural language and are based on a simple assumption:

> the probability of the next word in a sentence depends only on a fixed number of previous words

This assumption is the central aspect of n-gram models—the n being the number of previous words that the next word depends upon. Consider the sentence: "*He filled his coffee cup with* ___". When we read this sentence, we come up with many ideas for what the missing word might be—most of us might have guessed the next word to be "coffee" while some may guess "tea" and so on. This example illustrates the n-gram model: if the model has access to n previous words, it can use information from statistics of English to produce the next word based on likelihood in the corpora. Formally, for a sequence of words $w_1, \ldots, w_m$, the probability of observing these words in this sequence is given by:

$$P(w_1, \ldots, w_m) = \prod_{i=1}^{m} P(w_i | w_1, \ldots, w_{i-1}) \tag{2.1}$$

However, for an n-gram model, using the assumption above, this probability is approximated as:

$$P(w_1, \ldots, w_m) = \prod_{i=2}^{m} P(w_i | w_{i-(n-1)}, \ldots, w_{i-1}), \tag{2.2}$$

where the conditional probability is calculated as a fraction of counts from the text corpora that the statistical model is built on:

$$P(w_i | w_{i-(n-1)}, \ldots, w_{i-1}) = \frac{count(w_{i-(n-1)}, \ldots, w_{i-1}, w_i)}{count(w_{i-(n-1)}, \ldots, w_{i-1})} \tag{2.3}$$

n-gram models have been recently replaced by neural-network-based models that learn from large corpora of data. These neural models of natural language produce *word embeddings* that convert text into high-dimensional embeddings or vectors. We will learn about these techniques below.

2.2.2 Word Embeddings, Transformers, and Neural Language Models

To use neural models for various natural language processing tasks words need to be represented as vectors; such representation of words are referred to as word embeddings. The use of word embeddings in NLP pre-dates the almost universal use of neural networks in NLP these days. Earlier the use of word embeddings was referred to as distributional semantics [2]. One of their initial realization developed in the context of information retrieval was Latent Semantic Analysis (LSA) [3] or Latent Semantic Indexing (LSI) [4] where the vector representation of a word (or a term) first computes the number of times that word/term is present in a set of contextual documents.

The use of neural networks to build word embeddings was popularized by the Skipgram and CBoW models [5] of Word2Vec. Both Skipgram and CBoW (Continuous Bag Of Words) are two layered neural networks that take as input a 1-hot representation of words and optimize the conditional probability of a word given its neighboring word(s)—in the case of Skipgram it is the conditional probability of the context given a center word and in case of CBoW it is the opposite. Following is a brief description of the CBoW method.

Let V be the size of the vocabulary $w_1 \ldots w_V$ and N be the length of embedding vectors. Let $x^1 = [x_1^1, \ldots x_V^1], \ldots x^k = [x_1^k, \ldots x_V^k]$ be the 1-hot representation of k context words that are given as an input to a CBoW Word2Vec network that is defined as follows. $W \in R^{V \times N}$ and $W' \in R^{N \times M}$ are two matrices used in the network. $h^1 \ldots h^k$ are defined as $h^i = W^T x^i$ and $h = \frac{h^1 + \cdots + h^k}{k}$. $u \in R^{V \times 1}$ is defined as $W'^T h$, and $y = [x_1, \ldots y_v]$ is obtained by softmax on u. I.e.,

$$y_i = \frac{e^{u_i}}{\sum_{j=1}^{V} e^{u_j}}$$

Intuitively, y_i encodes the probability that w_i (the ith word in the vocabulary) is the center word given the context $x^1 \ldots x^k$. The matrices W and W' are learned by maximizing the conditional probability of center words given their contexts. Both matrices encode word embedding of size N for the words $w_1 \ldots w_V$ and usually W is used as the embedding for downstream tasks.

While Word2Vec and its variants such as GloVe were widely used for several years, there were two key issues of concern. First, in the computation of Word2Vec, the order of words is ignored in the computation $h = \frac{h^1 + \cdots + h^k}{k}$, and second, since words can have multiple meanings, and the correct meaning is often understood based on the context, a unique embedding of a word would not be able to capture the possibly multiple meanings of the word and thus embedding is needed that vary based on context.

Recurrent Neural Networks (RNNs) can embed words and sequences (of varied lengths) of words into vectors of fixed size. A generic RNN takes as an input a representation s_0 of an initial state vector and a sequence of words $w_1, \ldots w_n$ and computes the sequence of states $s_1, \ldots s_n$, using a recurrent function R as follows: $s_i = R(s_{i-1}, w_i)$. Associated with s_i is an output function O that is used to define the outputs $y_i = O(s_i)$. R and O

are implemented using neural networks and their parameters are learned by training the RNN on NLP task datasets that use the s_is and y_is to get to the output desired by the specific NLP task under consideration. LSTM (Long Short-Term Memory) is a particular kind of RNN where the states have a memory component and a hidden state component and the network defining R has gates. One of the motivations behind LSTM was to address the vanishing gradient problem associated with the simplistic implementations of RNN such as S-RNN where $R(s_{i-1}, w_i)$ was simply $g(w_i W^w + s_{i-1} W^s)$ with g being a non-linear activation function such as $tanh$ or $ReLU$. ELMO based on a bi-directional LSTM (bi-LSTM) computes embedings that vary based on the context. Its optimization function for training is a language model formulation of predicting the "next" word, but being bi-directional it combines two LSTMs, one in the forward direction and another in the backward direction, and jointly maximizes the "next" word prediction in both forward and backward directions.

RNNs have been used for various NLP tasks. For tasks such as machine translation, an encoder-decoder architecture that uses RNNs has been useful. A simple RNN-based encoder-decoder architecture has two RNNs, the first referred to as an encoder and the second referred to as the decoder, and the final state of the encoder part is given as an input to the state computation in each state of the decoder.

With a slight change in terminology, an encoder part can be defined by an RNN that takes an input sequence of words (of varied lengths) $w_1, \ldots, w_{T_w}$ and computes the sequence of hidden states $h_1, \ldots, h_{T_w}$, using a recurrent function R_1 as follows: $h_i = R_1(h_{i-1}, w_i)$. A final context vector c based on all the hidden vectors can be computed using a function h as follows: $c = q(\{h_1, \ldots, h_{T_w}\})$. As before, R_1 and q are computed using a neural network.

The decoder part predicts outputs $y_1, \ldots y_{t'}$ one by one using the context vector c as follows: $y_t = g(y_{t-1}, s_t, c)$ where $s_t = R_2(s_{t-1}, y_{t-1}, c)$ denotes the hidden state in the decoder part, and R_2 and g are computed using a neural network.

However, it was found that the best performing encoder-decoder RNNs have individual state outputs of the encoder part also used through a soft-search mechanism as input to the decoder state components. The intuition behind it is that during decoding it is better to also have the information about specific parts of the input sequence and those parts are identified through a soft-search mechanism referred to as "attention". More formally, $y_1, \ldots y_{t'}$ are now predicted using individualize context vectors c_ts as follows: $y_t = g(y_{t-1}, s_t, c_t)$ where $s_t = R_2(s_{t-1}, y_{t-1}, c_t)$, and the c_ts are defined using alignment terms $e_{tj} = a(s_{t-1}, h_j)$ in the following way:

$$c_t = \sum_{j=1}^{T_x} \alpha_{tj} h_j \text{ where } \alpha_{tj} = \frac{\exp(e_{tj})}{\sum_{k=1}^{T_x} \exp(e_{tk})}$$

In the above g, R_2 and a are implemented using neural networks. Intuitively, e_{tj} denotes how well the input around position j matches with the output at position t, and α_{tj} is a softmax version of that.

2.2.3 Transformer-Based Language Models

In the landmark paper "Attention is all you need" the authors proposed a novel architecture which they call a Transformer [6] that uses two kinds of attention and dispenses with the recurrence. It captures the notion of the meaning or representation of a word being influenced by the context (the surrounding words) through stacked multi-head self-attention, where each head can be intuitively thought of as a particular kind of influence. Multiple layers in the stack accumulate indirect attention in a transitive manner. For example, words/tokens w_i and w_j may not be directly related, but both of them may be related to the word/token w_k, and thus indirectly related. In such a case, the indirect relation will not be captured in the first layer but will be captured in the second layer.

To elaborate on the attention mechanism in a head, an input representation X of dimension $l \times d_e$ (where l denotes the input length of the words/tokens in the input, and d_e is the embedding size of each token) is projected to three components Q, K and V, referred to as query, key and value (in X) respectively; with each of their dimensions as $l \times d$, where $d = d_e/3$. Thus, corresponding to the ith head, $K_i = XW_i^K$, $Q_i = XW_i^Q$ and $V_i = XW_i^V$, where the dimensions of W_i^Q, W_i^K, and W_i^V are $d_e \times d$. Self-attention is defined between Q_i, K_i and V_i in the following way:

$$\text{head}_i = \text{Attention}(Q_i, K_i, V_i) = \text{softmax}\left(\frac{Q_i K_i^T}{\sqrt{d}}\right) V_i$$

The overall attention is then computed by concatenating the attention computed through each of the h heads in the following way, where W_O is an output projection matrix of dimension $h * d \times d_e$.

$$\text{MultiHead\_Attention}(X) = \text{Concat}(\text{head}_1, \ldots, \text{head}_h) W_O$$

To understand Attention(Q_i, K_i, V_i) a bit better let us look at the matrix representation of Q, K, QK^T, and $(QK^T)V$. Here we ignore the subscripts of Q. K and T, and use q_j, k_j, and v_j to denote the query, key, and value vector of the jth word/token.

$$Q = \begin{pmatrix} q_{1,1} & q_{1,2} & \cdots & q_{1,d} \\ q_{2,1} & q_{2,2} & \cdots & q_{2,d} \\ \vdots & \vdots & \ddots & \vdots \\ q_{l,1} & q_{l,2} & \cdots & q_{l,d} \end{pmatrix} = \begin{pmatrix} q_1 \\ q_2 \\ \vdots \\ q_l \end{pmatrix}$$

$$K^T = \begin{pmatrix} k_{1,1} & k_{2,1} & \cdots & k_{l,1} \\ k_{1,2} & k_{2,2} & \cdots & k_{l,2} \\ \vdots & \vdots & \ddots & \vdots \\ k_{1,d} & k_{2,d} & \cdots & k_{l,d} \end{pmatrix} = \begin{pmatrix} k_1 & k_2 & \cdots & k_l \end{pmatrix}$$

$$QK^T = \begin{pmatrix} q_1k_1 & q_1k_2 & \cdots & q_1k_l \\ q_2k_1 & q_2k_2 & \cdots & q_2k_l \\ \vdots & \vdots & \ddots & \vdots \\ q_lk_1 & q_lk_2 & \cdots & q_lk_l \end{pmatrix}$$

$$(QK^T)V = \begin{pmatrix} q_1k_1 & q_1k_2 & \cdots & q_1k_l \\ q_2k_1 & q_2k_2 & \cdots & q_2k_l \\ \vdots & \vdots & \ddots & \vdots \\ q_lk_1 & q_lk_2 & \cdots & q_lk_l \end{pmatrix} \begin{pmatrix} v_1 \\ v_2 \\ \vdots \\ v_l \end{pmatrix} = \begin{pmatrix} q_1k_1v_1 + q_1k_2v_2 + \cdots + q_1k_lv_l \\ q_2k_1v_1 + q_2k_2v_2 + \cdots + q_2k_lv_l \\ \vdots \\ q_lk_1v_1 + q_lk_2v_2 + \cdots + q_lk_lv_l \end{pmatrix}$$

In the above, each element of the $l \times l$ matrix QK^T is a dot product that measures the similarity between the query and key projections of the various words/tokens. Thus each q_ik_j in that matrix is a real number. $\frac{1}{d}$ is a scaling factor and the application of softmax to $\frac{QK^T}{\sqrt{d}}$ leads to a matrix where for each word/token (and its key) we have a probability distribution over how much that word/token is related to the other words/tokens. Thus attention$(Q_i, K_i, V_i) = \text{softmax}(\frac{Q_iK_i^T}{\sqrt{d}})V_i$ gives us a weighted partial embedding of each of the l words/tokens.

Figure 2.2 from Vaswani et al. [7] illustrates the overall architecture of a transformer. The left part of it is the encoder and the right part of it is the decoder. While encoder-decoder transformers are targeted toward tasks such as machine translation various transformer-based models have only one or the other. For example, BERT and its successors only have the encoder part, while OpenAIGPT and its successors only have the decoder part.

The input to the encoder is an embedding of a fixed size of tokens; where the tokens are obtained by a tokenizer that splits the textual input into smaller chunks. BERT used the Wordpiece tokenizer and had 37,000 tokens, while the OpenAI models used the BPE (Byte Pair Encoder) tokenizer Tiktoken. A positional encoding component is added to the token representations so that the same token at different positions has a different encoding. The resulting encoding is fed to a transformer layer, a key component of which is the multi-head attention that we discussed earlier. The dropout and normalization modules help in preventing overfitting and helping convergence respectively. The feed-forward network is simply a fully connected linear layer followed by a ReLU followed by another fully connected linear layer preserving the overall dimension and one purpose of this is to integrate the outputs of the multiple attention heads from just being a simple concatenation.

The decoder's input in a transformer with both an encoder and a decoder is different during test time from the training time. During test time the initial input (the input is referred to as "Outputs") to the decoder module is the start sequence symbol <SS>, and the predicted output token is added to the right in each iteration (this is referred to as auto-regression) until the predicted token is the End of the Sequence token or the maximum size of the input is reached. During training time the input is the ground truth target sequence prepended by

2.2 Language Modeling

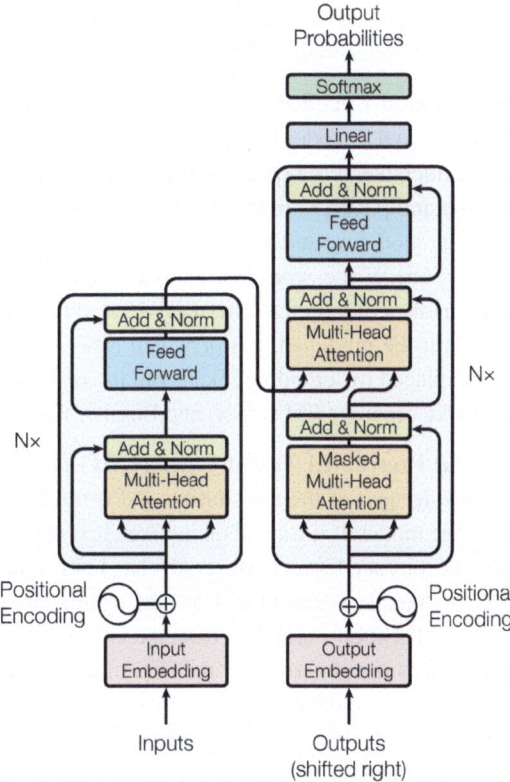

Fig. 2.2 The transformer model architecture. *Image credits* [7]

<SS> and it is trained to predict the target sequence. As evident from Fig. 2.2 there are three additional key differences between the decoder module and the encoder module:

(i) The multi-headed self-attention module is a masked one, where masking is done so that during the computation of attention at any position the information to the right is not visible;
(ii) there is a multi-headed cross attention module relating the encoder output (E) with the decoder's previous layer output (D); this is done by having $K_i = EW_i^K$, $Q_i = DW_i^Q$, $V_i = EW_i^V$; and
(iii) there is a linear layer followed by a softmax layer that would assign a probability to each token in the vocabulary and would be used for the prediction.

Built on the idea of transformers that we discussed above, several new methodologies for training language models have emerged in the last few years. These ideas have had a tremendous impact on natural language processing and resulted in the development of "*large language models*". The latest category of such large language models (such as ChatGPT) has had a much broader impact on discourse on artificial intelligence. We will briefly discuss some of these techniques below.

Bidirectional Encoder Representation from Transformers BERT [8], an acronym for, "Bidirectional Encoder Representation from Transformers" has only the encoder part of the transformer, has some special tokens (e.g., [SEP] to separate text sequences with a different purpose and [CLS] to indicate a classification task), and is trained on the tasks of masked language modeling and next sentence prediction. In masked language modeling the model predicts one of the words in a sequence, which may be masked or replaced by another random token or left intact. In BERT 15% of the input tokens during training are supposedly masked to be predicted, out of which 80% were replaced by the <MASK> token, 10% were replaced by a random token, and the rest 10% were left intact. BERT was pre-trained on BookCorpus (800M words) and English Wikipedia (2500 M words).

General Pretrained Transformer (GPT) GPT-2 [9] is a language model that only has the decoder part of the transformer and is trained on predicting the next word; making it a language model. It is trained on the WebText dataset, which contains 40 GB of data crawled from the internet. Its smallest version has 117 million parameters needing 500 MB of storage, while its largest version has 1,542 M parameters needing 6.5 GB of storage. To output a token GPT-2 allows sampling from the top k tokens and the model continues outputting until 1024 tokens are generated or the End-of-Sequence token is produced. In contrast, GPT-3, also a language model, has the token width of 2048, and is made of 96 decoder layers, and its self-attention layers are slightly different with alternating dense and sparse self-attention. It has 175 billion parameters and is trained on a dataset of 300 billion tokens of text consisting of CommonCrawl, WebText, Wikipedia, and a corpus of books. Its training is estimated to have cost 355 GPU years and $4.6 million.

The Emergency of Large Language Models BERT and GPT-2, among the earliest transformer-based language models, demonstrated rich language understanding, yet required supervised fine-tuning for specific downstream tasks like question-answering or summarization. GPT-3, however, marked a significant shift in this paradigm. With its 175 billion parameters, a stark increase from BERT's 220 million, and training on a vast corpus, GPT-3 introduced emergent abilities in zero-shot and few-shot learning. In zero-shot learning, the model performs unseen tasks based on user prompts, while in few-shot learning, it infers answers from a few contextual examples. The year 2022 witnessed a seminal moment with the introduction of ChatGPT, a model surpassing GPT-3 [10] by incorporating reinforcement learning with human feedback (RLHF) [11]. This advancement led to more refined and contextually aware responses, further ingraining language models in daily human activities. Today, ChatGPT and its successor, GPT-4, are ubiquitous in various domains, assisting with tasks ranging from email composition to code debugging and creative writing. Beyond their practical applications, these models also raise critical discussions about computational resource demands, ethical concerns like bias and misinformation, and the broader societal impact of AI.

2.2.4 Limitations of Large Language Models

Although the large language model (LLM) GPT-3 was impressive and surprised the NLP community and beyond, it faced the problem that at times it would misunderstand an input (referred to as a 'prompt') and generate text unintended by the user. (This is referred to as misalignment and making the LLMs produce output as intended by the prompt is referred to as alignment.) This is because often there are multiple reasonable completions of a prompt and although users may know what they want the LLM did not have a way to prefer one over the other. For example, the prompt "How to get a loan of $10,000" could be followed by sequences such as "with bad credit", "for a used car", or "from a bank", while the user may just have posed it as a generic question to get an answer. An initial approach to address this was to use supervised fine-tuning on the pre-trained models with pairs of the form (prompt, response). OpenAI used 13,000 human-created (prompt, response) pairs for InstructGPT [11]; and such fine-tuning is referred to as "instruction tuning" by many. Except for asking human-created instruction training data [12–14], there are also attempts to use language models to generate such training dsata [15, 16]. An approach that further improves the performance is allowing the ranking of various responses to a prompt, rather than assuming that a prompt has only one unique correct response. Reinforcement Learning with Human Preferences (RLHF) is such an approach where user preferences are solicited and converted to the form (prompt, winning_response, losing_response) and is used to train a reward model that gives a score to (prompt, response) pairs and the reward model is then used in a Reinforcement Learning (RL) framework to learn policies that maximize overall reward while not straying too far from the original LLM and the fine-tuned LLM. The RL framework for this defines states as prompts, actions as tokens generated by the LLM on the prompts, and the resulting state due to such an action as the concatenation of the generated token to the initial prompt. Finding an optimal policy—a mapping from states to actions—in such a framework corresponds to finding the optimal token response concerning a prompt. To scale the annotated data in instruction tuning and RLHF approaches such as self-instruct and RLAIF (RL from AI feedback) have been developed where the language models are used to generate (prompt, response) pairs and (prompt, winning_response, losing_response) triplets respectively.

2.3 Modeling Techniques for Vision

Effective image representation is a crucial component of multimodal information retrieval systems, as it enables the extraction of valuable features and information from visual data. The goal is to transform raw pixel data into a compact, meaningful, and structured representation that can be easily integrated with other modalities such as text.

Traditional image representation techniques, such as color histograms, texture descriptors, and local feature-based methods like Scale-Invariant Feature Transform (SIFT) [17] and Speeded-Up Robust Features (SURF) [18]. These methods, while effective in certain contexts, cannot often capture high-level semantic information and complex visual patterns. In this section, we will mainly discuss modern deep learning-based models.

Particularly, we will explore the advancements brought forth by deep learning, specifically Convolutional Neural Networks (CNNs) [19], which have revolutionized the field of computer vision. We will discuss the architecture and principles of CNNs and how they can automatically learn hierarchical and robust visual features from raw pixel data.

Next, we will describe ResNet [20], which is trained on massive datasets like ImageNet [21], has demonstrated exceptional performance in various computer vision tasks and can be fine-tuned to extract high-quality features for specific applications in multimodal information retrieval.

Lastly, we will discuss the emergence of transformer-based vision models, which have gained significant attention due to their ability to handle long-range dependencies and complex patterns in image data. These models adopt the transformer architecture, originally designed for natural language processing, to process image data by dividing it into a fixed number of non-overlapping patches and encoding them as a sequence of tokens. We will explore the principles behind these models, their advantages over traditional CNNs, and their applications in multimodal information retrieval systems.

2.3.1 CNN

CNNs are designed to effectively capture spatial patterns and hierarchical representations from visual data. The key building blocks of a CNN are convolutional layers, pooling layers, and fully connected layers.

- Convolutional layers: Convolutional layers are responsible for learning local patterns and features from the input data. They consist of multiple learnable filters or kernels that convolve over the input image, performing element-wise multiplication and summation to generate feature maps. Each filter specializes in detecting specific visual patterns, such as edges, corners, or textures. The sharing of weights across the input image allows CNNs to efficiently capture spatial information and achieve invariance to translations.
- Pooling layers: Pooling layers downsample the spatial dimensions of the feature maps, reducing the computational burden and extracting important features. The most common type of pooling is max pooling, which selects the maximum value within a given window. Pooling helps make the representation more compact, robust to small spatial translations, and less sensitive to minor variations in the input.
- Fully connected layers: Following the convolutional and pooling layers, fully connected layers are employed for high-level reasoning and decision-making. These layers connect every neuron from the previous layer to the subsequent layer, enabling the model to learn complex combinations of features and make predictions. The final fully connected layer

typically incorporates activation functions, such as softmax or sigmoid, to produce class probabilities or regression outputs.

CNNs are typically composed of multiple stacks of convolutional and pooling layers, followed by one or more fully connected layers. This arrangement allows the network to progressively learn complex and abstract representations of the input data. The depth of the network, i.e., the number of convolutional layers, plays a crucial role in capturing hierarchical features of varying complexity.

2.3.2 Residual Neural Network: ResNet

ResNet addresses the issue of vanishing gradients in very deep neural networks by utilizing residual connections. Residual blocks allow the network to learn residual mappings, which are the differences between the desired output and the input. By learning these residual mappings instead of the entire output, the network can focus on fine-tuning the residuals rather than trying to learn the complete mapping from scratch. A residual block typically consists of the following components,

- Input: The input to the residual block is a feature map or activation from the previous layer or block in the network.
- Convolutional Layers: The input passes through a series of convolutional layers within the residual block. These layers perform various operations such as convolution, batch normalization, and activation functions (commonly ReLU). The purpose of these layers is to capture and transform the input features.
- Shortcut or Skip Connection: A key component of the residual block is the skip connection, also known as the identity mapping. The skip connection directly propagates the input from one layer to a later layer within the block. This shortcut connection allows the gradient to flow directly through the network, facilitating easier learning of the residuals.
- Output: The output of the residual block is obtained by adding the transformed input features (output of the convolutional layers) and the input features passed through the skip connection. This element-wise addition combines the transformed features with the original input, effectively learning the residual mapping.

Residual blocks can be stacked together to create deeper ResNet architectures, where the output of one block becomes the input to the next block. The skip connections allow the gradients to propagate efficiently, facilitating the training of very deep networks with improved accuracy and performance.

2.3.3 Vision Transformers (ViT)

In recent years, there has been a significant shift in computer vision, driven by the remarkable success of convolutional neural networks (CNNs). However, Convolutional Neural Networks have limitations in terms of their global perception and their ability to handle long-range dependencies within an image. Vision Transformers (ViTs) have emerged as a revolutionary architecture that aims to address these limitations by leveraging the power of self-attention mechanisms, originally popularized in the context of natural language processing. In this section, we will introduce the fundamental concepts, architecture, and training procedures of Vision Transformers.

Introduction to Vision Transformers: The Vision Transformer (ViT) architecture, proposed by Dosovitskiy et al. [22] in 2020, presents a novel approach to process images by treating them as sequences of patches. This departure from the traditional grid-based processing allows ViTs to model global interactions between image regions more effectively, thus capturing both local and long-range dependencies.

At the core of the ViT model is the self-attention mechanism, specifically the multi-head self-attention mechanism introduced in the Transformer architecture. This mechanism enables ViTs to weigh the importance of different patch representations while considering their relationships, thereby enabling the model to capture rich contextual information.

Architecture of Vision Transformers: The ViT architecture can be divided into several key components:

- *Patch Embedding*: The input image is divided into non-overlapping patches, which are then linearly embedded into lower-dimensional vectors. These embeddings serve as the initial input to the subsequent layers of the ViT model.
- *Positional Embeddings*: Since ViT processes images as sequences of patches, it lacks inherent spatial information. To address this, positional embeddings are added to the patch embeddings, providing the model with a notion of patch positions within the image.
- *Transformer Encoder*: The heart of the ViT architecture lies in its Transformer encoder. This encoder consists of multiple layers, each comprising a multi-head self-attention mechanism and position-wise feedforward neural networks. The self-attention mechanism allows the model to attend to different patches based on their relevance to each other, capturing both local and global contextual information.

Mathematically, the multi-head self-attention mechanism can be defined as follows:

$$\text{MultiHead}(Q, K, V) = \text{Concat}(\text{head}_1, \ldots, \text{head}_h) W_O$$

where

$$\text{head}_i = \text{Attention}(Q W_{Qi}, K W_{Ki}, V W_{Vi})$$

and
$$\text{Attention}(Q, K, V) = \text{softmax}\left(\frac{QK^T}{\sqrt{d_k}}\right) V$$

Here, Q, K, and V represent the query, key, and value matrices, W_{Qi}, W_{Ki}, and W_{Vi} are learnable weight matrices, h is the number of attention heads, d_k is the dimension of the key vectors, and W_O is an output projection matrix.

- *Classification Head*: After processing the patch embeddings through the Transformer encoder, the final output representation is flattened and fed into a standard fully connected layer for classification. The number of output units in this layer corresponds to the number of target classes.

ViT applications: Vision Transformers have demonstrated exceptional performance across a range of computer vision tasks, including image classification, object detection, semantic segmentation, and more. Their ability to capture both local and global information has proven crucial for tasks that require understanding complex spatial relationships within images.

The Vision Transformer architecture represents a paradigm shift in computer vision, challenging the dominance of convolutional neural networks. By leveraging self-attention mechanisms and treating images as sequences of patches, ViTs have demonstrated state-of-the-art performance on various tasks. As research in this field continues to evolve, it is expected that Vision Transformers will play a pivotal role in shaping the future of computer vision applications.

2.4 Multimodal Models

In the field of Artificial Intelligence, the convergence of visual perception and linguistic comprehension has given birth to a new paradigm which become a hot field of research in the last decade: vision and language models (V&L models). These models represent a remarkable leap forward, enabling machines to simultaneously understand and interpret both visual information and linguistic context and use these cues to solve complex real-world tasks. By seamlessly blending these two modalities, integrated V&L models are reshaping the landscape of AI applications across a spectrum of industries.

Integrated V&L model models have emerged as a result of the convergence of advancements in both computer vision and natural language processing. These models leverage the power of convolutional neural networks (CNNs) based architectures for visual feature extraction and transformer-based architectures for linguistic understanding. Through joint training on multimodal datasets comprising images/videos and corresponding textual descriptions, these models learn to associate visual cues with linguistic semantics, effectively bridging the gap between visual and linguistic realms.

The research and development of integrated V&L models models has paved the way for a rich set of applications that rely on the synergy of perception and semantics. We list a few notable examples here:

- Image Captioning: Integrated models excel at generating descriptive captions for images. Given an image as input, the model analyzes the visual content and generates a coherent textual description that captures its essence. This technology finds applications in aiding visually impaired individuals, content indexing, and enhancing user experiences in photo-sharing platforms.
- Visual Question Answering (VQA): Integrated models enable machines to answer questions about images. By comprehending both the visual context and the posed question, these models provide accurate and contextually relevant responses. VQA has applications in educational tools, content retrieval, and robotics, where machines can interact with the environment based on visual cues.
- Visual Dialog Systems: These systems engage in dynamic conversations about images. Users can have interactive dialogues with machines by asking questions, providing prompts, and receiving detailed responses that combine both visual and linguistic understanding. This technology holds potential in customer service, virtual assistants, and educational platforms.
- Multi-Modal Retrieval: Integrated models facilitate the retrieval of information across modalities. Users can search for images using textual queries or find textual content based on visual inputs. This capability has implications in content recommendation, e-commerce, and information retrieval systems. In later chapters of the book, we will elaborate on many of these applications building upon multimodal retrieval settings.
- Medical Image Analysis: In the healthcare domain, integrated models can aid in diagnosing medical conditions by analyzing medical images and related clinical notes. This fusion of visual and textual data assists medical professionals in making accurate and informed decisions.

Integrated vision and language models exemplify the potential of AI to transcend traditional boundaries and simulate human-like understanding. As research continues to refine these models, their capabilities will undoubtedly expand, ushering in a new era of AI applications that seamlessly combine the power of visual perception and linguistic comprehension. In this chapter, we introduce a few notable integrated V&L architectures and their training paradigms.

2.4.1 LXMERT: Language Cross-Modal Pretraining for Vision-and-Language Understanding

There has been a growing interest in models that can understand both visual and textual information, enabling a deeper level of understanding and interaction between the two modalities. The Language Cross-Modal Pretraining for Vision-and-Language Understanding (LXMERT) model [23] is a pioneering architecture that addresses this challenge by jointly learning representations from both text and images. In this section, we introduce the foundational concepts, architecture, and training strategies of the LXMERT model.

The LXMERT model, introduced by Tan and Bansal in 2019, represents a significant advancement in the realm of vision and language understanding. It tackles the task of comprehending and generating textual descriptions for images, as well as answering questions about images, by leveraging the power of both modalities.

At its core, LXMERT combines concepts from natural language processing (NLP) and computer vision, facilitating a more comprehensive grasp of multimodal information. By jointly training on text and images, the model acquires a more holistic understanding of the relationships between visual and textual elements.

Architecture of LXMERT: The LXMERT architecture encompasses several key components that enable it to process and understand both text and image inputs:

- *Vision Encoder*: The visual input, usually an image or a set of images, is encoded using a convolutional neural network (CNN). This CNN transforms the raw pixel data into a high-level feature representation that captures salient visual information.
- *Text Encoder*: The textual input, such as a sequence of words or sentences, is encoded using a transformer-based architecture, similar to those used in natural language processing tasks. This text encoder captures the semantic meaning and relationships within the textual input.
- *Cross-Modal Encoder*: The cross-modal encoder integrates both the visual and textual encodings, facilitating the fusion of information from both modalities. This encoder leverages multimodal self-attention mechanisms to capture correlations between the visual and textual features, enabling the model to align and understand the associations between them.

Mathematically, the cross-modal self-attention mechanism can be described as follows:

$$\text{CrossModalAttention}(V, T) = \text{softmax}\left(\frac{V^T T}{\sqrt{d}}\right) T$$

Here, V represents the visual embeddings, T represents the textual embeddings, and d is the dimension of the embeddings.

- *Task-Specific Heads*: LXMERT is designed to perform various tasks, including image captioning, question answering, and more. To cater to these tasks, task-specific heads are attached to the cross-modal encoder. These heads are composed of fully connected layers that map the cross-modal embeddings to task-specific predictions.

Applications and Advancements LXMERT has found applications in a range of tasks such as visual question answering, image captioning, and referring expression comprehension. It has demonstrated impressive performance by effectively fusing visual and textual information. Advancements in the field have led to variations of LXMERT, including models that consider more complex interactions between modalities, improved architectures for more efficient training, and techniques for better handling of out-of-domain data.

2.4.2 OSCAR—Object-Semantics Aligned Pretraining for Vision-and-Language Understanding

The integration of vision and language has garnered significant attention due to its potential to enhance the understanding and interaction between these two modalities. The Object-Semantics Aligned Pretraining for Vision-and-Language Understanding (OSCAR) model stands at the forefront of this field, presenting a cutting-edge architecture that addresses the challenges of capturing rich semantic relationships between visual and textual information.

The OSCAR model, introduced by Li et al. [24], represents a significant leap forward in bridging the gap between vision and language. By aligning objects in images with corresponding semantic concepts in text, OSCAR achieves a more nuanced understanding of multimodal data. This alignment allows the model to capture intricate relationships that are essential for accurate vision and language understanding.

Architecture of OSCAR
The OSCAR architecture is built upon several crucial components that enable seamless processing and alignment of both visual and textual information:

- *Object Detection*: Incorporating object detection within the model, OSCAR identifies and localizes objects within images. This step not only enhances the model's visual understanding but also establishes a direct link between visual objects and textual concepts.
- *Textual Encoder*: The textual input, typically consisting of sentences or paragraphs, is encoded using a transformer-based architecture. This encoder captures the semantics of the textual data, allowing OSCAR to grasp the nuances of language.
- *Object-Text Alignment*: The key innovation of OSCAR lies in its alignment mechanism, which associates visual objects with corresponding textual concepts. This alignment ensures that the model captures the relationships between objects and concepts, enabling a richer multimodal representation.

2.4 Multimodal Models

Mathematically, the alignment mechanism can be described as:

$$\text{Alignment}(V, T) = \text{softmax}\left(\frac{V^T T}{\sqrt{d}}\right) T$$

Here, V represents visual embeddings, T represents textual embeddings, and d is the dimension of the embeddings.

- *Cross-Modal Encoder*: Following alignment, a cross-modal encoder combines visual and textual features, facilitating the fusion of information from both modalities. Multimodal self-attention mechanisms enable OSCAR to capture intricate correlations between visual objects and textual concepts.
- *Task-Specific Heads*: To perform specific tasks like image captioning or visual question answering, OSCAR employs task-specific heads that transform cross-modal embeddings into task predictions. These heads comprise fully connected layers that tailor the model's output to the specific requirements of each task.

Applications and Advancements: OSCAR's innovative approach to aligning objects with semantics has led to exceptional performance in various vision-and-language tasks, including image-text matching, image captioning, and more. Researchers have built upon OSCAR's foundation, proposing advancements that refine alignment mechanisms, enhance model architectures, and expand its applicability to different domains.

2.4.3 ViLT—Vision and Language Transformer

The convergence of vision and language has become a pivotal research area, enabling models to comprehend and generate information from both modalities. The Vision and Language Transformer (ViLT) model stands as a remarkable advancement in this domain, seamlessly integrating visual and textual information for enhanced multimodal understanding.

The Vision and Language Transformer (ViLT), introduced by Kim et al. [25], represents a breakthrough in the fusion of vision and language within a single architecture. By jointly modeling visual and textual data, ViLT transcends the limitations of unimodal models and provides a holistic understanding of multimodal content.

Architecture of ViLT: ViLT's architecture encompasses critical components that facilitate the integration of visual and textual modalities:

- *Visual Feature Extraction*: Images are processed through convolutional neural networks (CNNs) to extract high-level visual features. These features provide a rich representation of visual content, forming the foundation for multimodal fusion.
- *Textual Embedding*: Textual input, comprising sentences or paragraphs, undergoes embedding via transformer-based architectures. This textual embedding captures the

semantic nuances of the language, allowing ViLT to understand textual information effectively.
- *Multimodal Transformer*: ViLT's hallmark lies in its Multimodal Transformer, a variant of the transformer architecture adapted to handle both visual and textual data. This architecture facilitates cross-modal attention mechanisms, enabling ViLT to align visual features with corresponding textual embeddings. Mathematically, the cross-modal attention can be described as follows:

$$\text{CrossModalAttention}(V, T) = \text{softmax}\left(\frac{V^T T}{\sqrt{d}}\right) T$$

Here, V represents visual embeddings, T represents textual embeddings, and d is the embedding dimension.
- *Task-Specific Heads*: To perform various tasks such as image captioning or visual question answering, ViLT employs task-specific heads. These heads, comprising fully connected layers, map the fused multimodal embeddings to task-specific predictions.

Applications and Advancements: ViLT's pioneering approach to vision and language integration has led to impressive results across various tasks, including image-text retrieval, captioning, and more. Researchers continue to advance the ViLT architecture, exploring more intricate fusion mechanisms, novel attention strategies, and optimizations for performance and efficiency.

2.4.4 CLIP—Contrastive Language-Image Pretraining

In multimodal learning, the CLIP (Contrastive Language-Image Pretraining) model stands as a groundbreaking advancement, enabling a joint understanding of images and natural language. Developed by Radford et al. [26], CLIP transcends traditional boundaries by learning to associate images and text in a truly cross-modal manner.

CLIP represents a paradigm shift in how models approach multimodal understanding. Rather than relying solely on curated datasets, CLIP learns to align images and text through a self-supervised approach. By enabling a shared embedding space for images and text, CLIP enables a deeper level of comprehension that extends across both modalities.

Architecture of CLIP: The CLIP model comprises key components that facilitate its cross-modal understanding:

- *Vision Encoder*: Images are processed through a convolutional neural network (CNN) to extract high-level visual features. These features form the basis for cross-modal alignment.

- *Textual Encoder*: Text inputs, such as descriptions or captions, are encoded using transformer-based architectures. This process captures the semantic nuances of language for cross-modal understanding.
- *Contrastive Loss*: CLIP's innovation lies in its use of a contrastive loss function. This loss encourages positive pairs (image-text pairs) to be closer in the embedding space while pushing negative pairs (image-text pairs from different sources) apart. This way, CLIP learns to align corresponding images and text, making the association more robust. Mathematically, the contrastive loss can be expressed as:

$$\mathcal{L} = -\log \frac{\exp(\text{sim}(I, T))}{\sum_N \exp(\text{sim}(I, T_N))}$$

Here, I represents image embeddings, T represents text embeddings, N denotes negative pairs, and $\text{sim}(\cdot, \cdot)$ computes the similarity between embeddings.
- *Joint Embedding Space*: CLIP's ultimate achievement is a joint embedding space where corresponding image and text pairs are close, while dissimilar pairs are distant. This space enables the model to perform tasks across both modalities seamlessly.

Training CLIP: Training CLIP involves a contrastive learning setup:

- *Negative Sampling*: For each image, negative samples (textual descriptions from different sources) are selected. These negative samples encourage the model to distinguish between corresponding and non-corresponding pairs.
- *Pretraining*: During pretraining, CLIP learns to align images and text using contrastive loss. This step equips the model with cross-modal understanding without the need for task-specific annotations.

Applications and Advancements: CLIP's revolutionary approach has propelled it to excel in a wide array of tasks, including image classification, object detection, and even zero-shot image generation from textual prompts. Researchers continue to build on CLIP's foundation, exploring variations, scaling strategies, and fine-tuning for specific tasks.

2.5 Broader Impact of Transformer-Driven Models for Language, Vision, and Multimodality

The advent and evolution of transformer-driven models have precipitated a paradigm shift in the fields of language processing, computer vision, and multimodal interactions, with far-reaching impacts across technology, academia, and society. Technologically, these models have dramatically enhanced natural language processing capabilities, evidenced by the sophistication of chatbots like OpenAI's GPT-4. In computer vision, they have revolutionized image recognition and analysis. Academically, transformers have catalyzed interdisciplinary research, creating a nexus between linguistics, neuroscience, and computer science. This has opened new research avenues, deepening our understanding of both artificial and human intelligence. The societal impacts are profound, particularly in enhancing accessibility technologies for the differently-abled, such as AI-driven assistive devices that translate text to speech for the visually impaired.

The influence of Transformer models has notably expanded into the realm of multimodal applications as well. Integrating visual and textual data, transformers have enabled more responsive, adaptive, and contextually aware systems. These advancements are especially pronounced in the domain of multimodal retrieval, where transformer-based models facilitate efficient and accurate retrieval of relevant information from diverse datasets, significantly enhancing user experience in applications like search engines and recommendation systems.

However, the technology is not without limitations. The data-hungry nature of these models and the quest for robust, generalizable AI remains a significant challenge. This transformative technology not only redefines current computational methodologies but also holds the promise of significantly impacting our societal, academic, and professional landscapes in the years to come, provided we navigate its ethical, environmental, and global implications responsibly.

In the following chapters, we will learn about the task of multimodal retrieval and how this task utilizes many of the ideas and models for language representation and visual representation that we discussed in this chapter.

References

1. Dan Jurafsky. *Speech & language processing*. Pearson Education India, 2000.
2. Gemma Boleda. Distributional semantics and linguistic theory. *Annual Review of Linguistics*, 6:213–234, 2020.
3. Susan T Dumais. Latent semantic analysis. *Annual Review of Information Science and Technology (ARIST)*, 38:189–230, 2004.
4. Susan Dumais et al. Latent semantic indexing (lsi) and trec-2. *Nist Special Publication Sp*, pages 105, 1994.
5. Tomas Mikolov, Kai Chen, Greg Corrado, and Jeffrey Dean. Efficient estimation of word representations in vector space. *arXiv preprint* arXiv:1301.3781, 2013.
6. Ashish Vaswani, Noam Shazeer, Niki Parmar, Jakob Uszkoreit, Llion Jones, Aidan N. Gomez, Lukasz Kaiser, and Illia Polosukhin. Attention is all you need. In Isabelle Guyon, Ulrike von Luxburg, Samy Bengio, Hanna M. Wallach, Rob Fergus, S. V. N. Vishwanathan, and Roman Garnett, editors, *Advances in Neural Information Processing Systems 30: Annual Conference on Neural Information Processing Systems 2017, December 4–9, 2017, Long Beach, CA, USA*, pages 5998–6008, 2017b. URL https://proceedings.neurips.cc/paper/2017/hash/3f5ee243547dee91fbd053c1c4a845aa-Abstract.html.
7. Ashish Vaswani, Noam Shazeer, Niki Parmar, Jakob Uszkoreit, Llion Jones, Aidan N. Gomez, Lukasz Kaiser, and Illia Polosukhin. Attention is all you need. In Isabelle Guyon, Ulrike von Luxburg, Samy Bengio, Hanna M. Wallach, Rob Fergus, S. V. N. Vishwanathan, and Roman Garnett, editors, *Advances in Neural Information Processing Systems 30: Annual Conference on Neural Information Processing Systems 2017, December 4–9, 2017, Long

Beach, CA, USA, pages 5998–6008, 2017a. URL https://proceedings.neurips.cc/paper/2017/hash/3f5ee243547dee91fbd053c1c4a845aa-Abstract.html.

8. Jacob Devlin, Ming-Wei Chang, Kenton Lee, and Kristina Toutanova. BERT: Pre-training of deep bidirectional transformers for language understanding. In *Proceedings of the 2019 Conference of the North American Chapter of the Association for Computational Linguistics: Human Language Technologies, Volume 1 (Long and Short Papers)*, pages 4171–4186, Minneapolis, Minnesota, 2019a. Association for Computational Linguistics. https://doi.org/10.18653/v1/N19-1423. URL https://aclanthology.org/N19-1423.

9. Tianyu Gao, Adam Fisch, and Danqi Chen. Making pre-trained language models better few-shot learners. In *Proceedings of the 59th Annual Meeting of the Association for Computational Linguistics and the 11th International Joint Conference on Natural Language Processing (Volume 1: Long Papers)*, pages 3816–3830, Online, 2021a. Association for Computational Linguistics. https://doi.org/10.18653/v1/2021.acl-long.295. URL https://aclanthology.org/2021.acl-long.295.

10. Partha Pratim Ray. Chatgpt: A comprehensive review on background, applications, key challenges, bias, ethics, limitations and future scope. *Internet of Things and Cyber-Physical Systems*, 2023.

11. Long Ouyang, Jeffrey Wu, Xu Jiang, Diogo Almeida, Carroll Wainwright, Pamela Mishkin, Chong Zhang, Sandhini Agarwal, Katarina Slama, Alex Ray, et al. Training language models to follow instructions with human feedback. *Advances in Neural Information Processing Systems*, 35:27730–27744, 2022.

12. Swaroop Mishra, Daniel Khashabi, Chitta Baral, and Hannaneh Hajishirzi. Cross-task generalization via natural language crowdsourcing instructions. In *Proceedings of the 60th Annual Meeting of the Association for Computational Linguistics (Volume 1: Long Papers)*, pages 3470–3487, 2022.

13. Yizhong Wang, Swaroop Mishra, Pegah Alipoormolabashi, Yeganeh Kordi, Amirreza Mirzaei, Atharva Naik, Arjun Ashok, Arut Selvan Dhanasekaran, Anjana Arunkumar, David Stap, et al. Super-naturalinstructions: Generalization via declarative instructions on 1600+ nlp tasks. In *Proceedings of the 2022 Conference on Empirical Methods in Natural Language Processing*, pages 5085–5109, 2022a.

14. Mihir Parmar, Swaroop Mishra, Mirali Purohit, Man Luo, Murad Mohammad, and Chitta Baral. In-BoXBART: Get instructions into biomedical multi-task learning. In *Findings of the Association for Computational Linguistics: NAACL 2022*, pages 112–128, Seattle, United States, 2022. Association for Computational Linguistics. https://doi.org/10.18653/v1/2022.findings-naacl.10. URL https://aclanthology.org/2022.findings-naacl.10.

15. Yizhong Wang, Yeganeh Kordi, Swaroop Mishra, Alisa Liu, Noah A Smith, Daniel Khashabi, and Hannaneh Hajishirzi. Self-instruct: Aligning language model with self generated instructions. *arXiv preprint* arXiv:2212.10560, 2022b.

16. Rohan Taori, Ishaan Gulrajani, Tianyi Zhang, Yann Dubois, Xuechen Li, Carlos Guestrin, Percy Liang, and Tatsunori B Hashimoto. Stanford alpaca: An instruction-following llama model, 2023.

17. David G Lowe. Object recognition from local scale-invariant features. In *Proceedings of the seventh IEEE international conference on computer vision*, volume 2, pages 1150–1157. Ieee, 1999.

18. Herbert Bay, Tinne Tuytelaars, and Luc Van Gool. Surf: Speeded up robust features. In *Computer Vision–ECCV 2006: 9th European Conference on Computer Vision, Graz, Austria, May 7–13, 2006. Proceedings, Part I 9*, pages 404–417. Springer, 2006.

19. Yann LeCun, Léon Bottou, Yoshua Bengio, Patrick Haffner, et al. Gradient-based learning applied to document recognition. *Proceedings of the IEEE*, 86(11):2278–2324, 1998.

20. Kaiming He, Xiangyu Zhang, Shaoqing Ren, and Jian Sun. Deep residual learning for image recognition. In *2016 IEEE Conference on Computer Vision and Pattern Recognition, CVPR 2016, Las Vegas, NV, USA, June 27–30, 2016*, pages 770–778. IEEE Computer Society, 2016. https://doi.org/10.1109/CVPR.2016.90. URL https://doi.org/10.1109/CVPR.2016.90.
21. Jia Deng, Wei Dong, Richard Socher, Li-Jia Li, Kai Li, and Fei-Fei Li. Imagenet: A large-scale hierarchical image database. In *2009 IEEE Computer Society Conference on Computer Vision and Pattern Recognition (CVPR 2009), 20–25 June 2009, Miami, Florida, USA*, pages 248–255. IEEE Computer Society, 2009. https://doi.org/10.1109/CVPR.2009.5206848.
22. Alexey Dosovitskiy, Lucas Beyer, Alexander Kolesnikov, Dirk Weissenborn, Xiaohua Zhai, Thomas Unterthiner, Mostafa Dehghani, Matthias Minderer, Georg Heigold, Sylvain Gelly, et al. An image is worth 16x16 words: Transformers for image recognition at scale. *arXiv preprint arXiv:2010.11929*, 2020.
23. Hao Tan and Mohit Bansal. LXMERT: Learning cross-modality encoder representations from transformers. In *Proceedings of the 2019 Conference on Empirical Methods in Natural Language Processing and the 9th International Joint Conference on Natural Language Processing (EMNLP-IJCNLP)*, pages 5100–5111, Hong Kong, China, 2019a. Association for Computational Linguistics. https://doi.org/10.18653/v1/D19-1514. URL https://aclanthology.org/D19-1514.
24. Xiujun Li, Xi Yin, Chunyuan Li, Pengchuan Zhang, Xiaowei Hu, Lei Zhang, Lijuan Wang, Houdong Hu, Li Dong, Furu Wei, et al. Oscar: Object-semantics aligned pre-training for vision-language tasks. In *European Conference on Computer Vision*, pages 121–137. Springer, 2020.
25. Wonjae Kim, Bokyung Son, and Ildoo Kim. Vilt: Vision-and-language transformer without convolution or region supervision. In Marina Meila and Tong Zhang, editors, *Proceedings of the 38th International Conference on Machine Learning, ICML 2021, 18–24 July 2021, Virtual Event*, volume 139 of *Proceedings of Machine Learning Research*, pages 5583–5594. PMLR, 2021. URL http://proceedings.mlr.press/v139/kim21k.html.
26. Alec Radford, Jong Wook Kim, Chris Hallacy, Aditya Ramesh, Gabriel Goh, Sandhini Agarwal, Girish Sastry, Amanda Askell, Pamela Mishkin, Jack Clark, Gretchen Krueger, and Ilya Sutskever. Learning transferable visual models from natural language supervision. In Marina Meila and Tong Zhang, editors, *Proceedings of the 38th International Conference on Machine Learning, ICML 2021, 18–24 July 2021, Virtual Event*, volume 139 of *Proceedings of Machine Learning Research*, pages 8748–8763. PMLR, 2021. URL http://proceedings.mlr.press/v139/radford21a.html.

Multimodal Information Retrieval

In today's rapidly evolving digital landscape, the wealth of available information has expanded beyond the boundaries of traditional text-based content. With the proliferation of multimedia platforms and data sources, we are constantly bombarded with a rich variety of images, videos, audio, and text. This vast array of heterogeneous data poses new challenges and opportunities for the field of Information Retrieval (IR). To address these challenges and harness the potential of multimodal information, researchers and practitioners have turned their attention toward the development of Multimodal Information Retrieval (**MMIR**) systems.

IR systems of the past focused on unimodal retrievals, such as retrieving text documents relevant to text queries or finding images similar to image queries. Cross-modal retrieval has also been explored to develop functionalities such as retrieving relevant images corresponding to a text query, or retrieving relevant text documents corresponding to an image query. In unimodal retrieval, both input and output modalities are the same, and thus the methods, functions, or encoders to represent query and target can be the same. On the other hand, cross-modal retrieval has to use different functions to represent the query and source. In both unimodal and cross-modal retrieval, either the query or the target is represented using a single modality. More recently, the concept of multimodal information retrieval has emerged, where the query is represented by multiple modalities, such as a combination of images and text.

The unique perspective of this type of retrieval is that to capture the information needed from the query so that it can retrieve the right target, the query function needs to comprehend both the modalities and fuse information. In other words, if a system only looks at a single modality, then it ignores the key information present in the other modality and thus has a lower chance of retrieving the ground truth targets. Take, for example, a scenario where you want to identify a place you visited in New York City, based on a photo featuring various landmarks and people. A text-only retrieval system would fall short as it lacks visual context, while an image-only system might inaccurately focus on irrelevant elements in the photo. In contrast, a well-designed MMIR system allows you to input the phrase *"Where is this place?"* along with the photo, enhancing the accuracy of the search by combining textual and visual data. With the above motivation, this chapter explores the intricacies of MMIR and provides a comprehensive understanding of the techniques and approaches employed in the field with the following structure.

We will begin by introducing the basic concept of IR systems which will lay the foundation for understanding the mechanism of IR. In this section, we will cover the concepts of query and target, indexing, and scoring functions. Then, we describe the state-of-the-art retrieval models for unimodal and multimodal IR systems. The unimodal retrieval is the foundation of multimodal IR including the text and the image IR. In the section on Multimodal IR, we will differentiate it with the cross-modal IR, and focus on multimodal-query IR. We will discuss two representative multimodal-query IR in detail. After this, we will discuss applications application of multimodal IR in crucial downstream tasks. Later, we will discuss the evaluation metrics spanning from traditional evaluation to advanced semantic-based measurement. Finally, we will discuss the broader impact of MMIR.

3.1 Multimodal Data and Multimodal Learning

Multimodal data refers to data that is collected from multiple sources or modalities, such as text, images, audio, video, sensor data, and more. In other words, it involves information from different sensory channels or types of data that can be used together to gain a more comprehensive understanding of a particular phenomenon, event, or context. We live in a multimodal information environments, some common examples include:

- Combining textual descriptions with accompanying images or videos, commonly found on social media platforms, e-commerce websites, and in medical diagnostics.
- Analyzing spoken language (audio) alongside transcriptions (text) to improve automatic speech recognition, sentiment analysis, and voice assistants.
- Integrating sensor data (e.g., GPS, accelerometer) with video footage for applications in autonomous vehicles, surveillance, and sports analytics.
- Combining various biometric modalities like fingerprints, facial recognition, and voiceprints for authentication and security purposes.

Multimodal research plays a pivotal role by leveraging various data types to address complex issues, develop advanced technologies, and deepen our understanding of the world. This approach is particularly significant in areas like artificial intelligence (AI), computer vision (CV), natural language processing (NLP), and human-computer interaction (HCI). The advantages of multimodal research over single-modal approaches include a richer, more nuanced grasp of intricate phenomena, enabling more fluid and intuitive human-computer interactions, and improving perception and decision-making by integrating diverse sensor inputs like cameras, LiDAR, and radar. In healthcare, this methodology shows great promise; amalgamating patient records, medical images, sensor data, and genetic information, paves the way for more tailored and effective treatments. When it comes to developing applications, multimodal retrieval presents unique challenges, especially in the fusion and alignment of different modalities, which we will explore in more detail in subsequent sections.

3.1.1 Multimodality Fusion

The main goal of multimodality fusion is to combine features or information from different modalities to make a joint decision or prediction. Integrating image and text features can enhance the system's ability to capture richer semantic relationships, as each modality may provide different perspectives and insights on the underlying content. For example, images can convey visual patterns, colors, and spatial relationships, while text can provide explicit descriptions, contextual information, and high-level abstractions. By combining these features, the retrieval system can better understand the content and make more accurate and relevant decisions. We will begin by discussing early fusion techniques, which involve combining features at the representation level before feeding them into a joint model. This can be achieved through methods such as concatenation, element-wise addition, or element-wise multiplication of the features. We will then explore late fusion techniques, which combine the outputs of separate modality-specific models to generate a final decision. This can be accomplished using methods like weighted averaging, voting, or learning a fusion function. Next, we will discuss more advanced and sophisticated fusion methods, such as attention mechanisms, which can dynamically weigh and combine features from different modalities based on their relevance to the task at hand. We will also discuss the use of transformer-based models for multimodal feature fusion, as they provide a unified architecture that can handle both image and text data effectively.

Late Fusion. In late fusion, both modalities are processed independently through separate neural networks to extract features, such as using CNN, RNN, and transformer-based models. The extracted features from both modalities are then concatenated or combined in a later stage, often using fully connected layers, before making a final prediction or inference. After feature extraction, the outputs of these separate models are combined. This can be done through various methods like simple concatenation, weighted averaging, or more complex operations using a neural network layer. This approach allows the models to specialize in processing their respective modalities before combining them.

Early Fusion. Early fusion involves combining text and image information at the input level. The text and image data are fed into the neural network together, allowing the model to jointly learn representations that capture the interactions between the modalities from the beginning. This fusion can be achieved by concatenating or stacking the textual and visual input data before passing them through the network. Early fusion can promote richer interactions between modalities but may require careful design to handle input size discrepancies.

Cross-Modal Attention. Cross-modal attention mechanisms enable the model to attend to relevant parts of one modality based on the information from the other modality. These mechanisms allow the model to dynamically focus on specific regions or words that are most informative for fusion. For example, in an image captioning task, an attention mechanism can be used to attend to specific image regions while generating the textual description. Similarly, in visual question answering (VQA), attention can be used to align relevant image regions with the words in the question.

Multimodal Transformers. Transformer-based architectures, which have been successful in language and vision tasks, are also employed for text-image fusion. They allow the modeling of interactions between text and image modalities using self-attention mechanisms. The transformer's attention mechanisms can capture the relationships between words and image regions, enabling cross-modal fusion at various layers of the architecture. Multimodal transformers facilitate the capture of long-range dependencies and improve the integration of text and image information.

3.1.2 Multimodality Alignment

The main goal of Multimodality Alignment is to align or map features from different modalities into a common representation space. In this space, similar features (regardless of their original modality) will be close, and dissimilar features will be far apart. Properly aligning modalities can help a model to better understand the intricate relationships between them. This understanding can lead to more accurate and richer representations of data, which in turn can significantly improve the performance of tasks like multimodal retrieval tasks. The main challenge of multimodal alignment is to close the semantic gap between different modalities. Different modalities inherently represent information in different ways. For instance, the emotion "happiness" might be expressed in text as the word "happy", in images via a smiling face, and in audio through a certain tone of voice. There are two major types of modality alignment: global and local alignment.

Global Alignment. This type of alignment focuses on matching different modalities from a global perspective. For instance, it seeks to align the overall features of an image with those of a corresponding caption. An illustrative example of global alignment can be found in the CLIP model [1].

Local Alignment. Local alignment, on the other hand, concentrates on matching sub-objects or specific components of different modalities. For example, when given an image along with its corresponding caption, the local alignment would aim to align the local image

3.1 Multimodal Data and Multimodal Learning

feature representing a "man" with the local text feature describing "man". An instance of local alignment can be seen in the phrase-grounding pretraining of the GLIP model [2].

Learning a joint embedding is the major method for multimodal alignment. The goal of joint embeddings is to project the extracted features from different modalities into a shared embedding space where semantically similar items, regardless of their original modality, are close to each other. First, for each modality, features are extracted. As an example, BERT can be used to extract the language feature and ViT can be used to extract the image feature. The extracted features are then projected into a shared embedding space. Contrastive learning is widely used to learn the joint embeddings and we will discuss this in detail in the following.

Constrastive Learning is widely used in retrieval systems, such as CLIP [1] and DPR [3]. The contrastive loss function encourages the model to minimize the distance (similarity) between embeddings of positive pairs while maximizing the distance between embeddings of negative pairs. Two common formulations of contrastive loss are the triplet loss and InfoNCE (Noise-Contrastive Estimation) loss.

- Triplet Loss: It involves three embeddings: an anchor, a positive example, and a negative example. The loss is minimized when the distance between the anchor and the positive example is smaller than the distance between the anchor and the negative example by a certain margin.

$$\mathcal{L}_{\text{triplet}}(A, P, N) = \max(0, d(A, P) - d(A, N) + \alpha), \tag{3.1}$$

where A represents the anchor embedding, P represents the positive example embedding (similar example), N represents the negative example embedding (dissimilar example), $d(A, P)$ denotes the distance (e.g., Euclidean distance or cosine similarity) between the anchor and positive example embeddings, $d(A, N)$ denotes the distance between the anchor and negative example embeddings, and α is a margin, a hyperparameter that specifies how far apart the anchor and positive example should be compared to the anchor and negative example.

- InfoNCE Loss: This is a generalization of the triplet loss that uses a batch of examples and encourages the model to distinguish between positive and negative pairs in the batch.

$$\mathcal{L}_{\text{InfoNCE}} = -\log\left(\frac{\exp(\text{sim}(A, P))}{\exp(\text{sim}(A, P)) + \sum_{i=1}^{K} \exp(\text{sim}(A, N_i))}\right), \tag{3.2}$$

where sim is a similarity score function (e.g., cosine similarity or inner-dot product), K is the number of negative examples in the set, other notations are the same as those in the Triplet loss. In-batch negative is usually combined with InfoNCE loss for the sake of efficiency. In this process, negative examples are selected from within the same mini-batch of data that are used for training. This means that for each anchor example in the batch, other examples from the same batch are picked as the negative examples. These examples are part of the current training batch and are readily available for computing gradients during training.

Chen et al. [4] proposes SimCLR to learn visual representations, and they find the data augmentations (such as stochastic data augmentation, cropping, color distortion, and Gaussian Blur), adding a learnable nonlinear transformation between the representation and the contrastive loss, longer training time, and larger batch size can be beneficial for the training.

3.2 Basic Elements of IR Systems

The objective of Information Retrieval (IR) systems is to identify and retrieve the most pertinent items from a data collection in response to a user's query. Constructing an IR system involves four key elements: the query, the target, the indexing, and the scoring function.

3.2.1 Query

Queries are user-initiated requests to obtain specific information. Commonly, queries take the form of free-text questions like "What is the world's largest city?" However, they don't always have to be questions; even a single keyword can serve as a query. For instance, if a patient comes across the term "metastasis" in a clinical note and wants to understand it, the term "metastasis" can be the query. In today's digital era, with the ubiquity of mobile phones, images can also act as queries. People can effortlessly snap a photo and use it to seek information. Next, we discuss unimodal queries and multimodal queries and the corresponding benchmarks.

Unimodal Query. Unimodal Query means that the query is represented by only one type of data, such as text-only or image-only. The goal of multimodal retrieval with unimodal query is to use one type of data as a query to retrieve another type of data (e.g. image) as a target. The most popular benchmarks of unimodal queries include MS-COCO [5], Flick30K [6] and Recipe1M+ [7]. The latter two can be taken as text-to-image retrieval or image-to-text retrieval.

Multimodal Query. A multimodal query involves using multiple data types to formulate a query. Sometimes, a single data format such as text may not suffice to relay all required information effectively. For example, if someone sees a flower and wants to find where to buy it, a query using only text is limited, especially without knowing the flower's name. In contrast, a multimodal query that combines an image of the flower captured with a mobile camera and a text phrase like "shops that sell this flower" can provide the necessary detail. To advance research in this field, a range of benchmarks has been developed, including OkVQA [8], A-OKVQA [9], WebQA [10], ReMuQ [11], Fashion200k [12], and FashionIQ [13].

3.2.2 Target

Targets are the primary units or items that the system aims to retrieve which form a corpus or collection. The majority types of corpus include free-form text such as a passage in Wikipedia, images such as Google images, multimodal documents which are represented by different modalities, and knowledge graphs such as Wikidata and ConceptNet. Targets are typically stored semantically (i.e., using natural language symbols that can be understood by humans) in a structured form of knowledge bases or as free-form text articles such as books, news articles, encyclopedias, etc. Knowledge can also be stored as graphs, in the form of pre-trained generative models trained on large amounts of data that can distill knowledge implicitly in the form of neural network parameters. Some knowledge may also be contextual—for instance, in V&L navigation tasks, the relative location of objects in a room can differ for each environment but is crucial in solving the navigation problem. Below, we summarize the sources of external knowledge for V&L.

Knowledge Graphs. Common examples include ConceptNet, WordNet, DBpedia, Freebase, ATOMIC, hasPartKB, and WebChild. Commonsense knowledge is easily found in these knowledge bases, especially in ConceptNet. This knowledge is presented by graphical structure or a set of (subject, predicate, object) triples, and thus requires a graph embedding [14, 15]. Knowledge graphs have some limitations, (1) they are usually hand-written by humans and will not evolve while world knowledge gets updated frequently, (2) knowledge coverage is limited since much knowledge is hard to be represented by the triplets and many domains do not have knowledge bases. VisualGenome [16] is a structured knowledge base for images that connect objects in an image with attributes, region descriptions, and relationships with other objects, together in a "scene graph". Each component of the scene graph is mapped to a WordNet synset [17]. Alberts et al. [18] introduce VisualSem, a multilingual and multimodal KG. VisualSem is linked to Wikipedia articles, WordNet synset [17], and images from ImageNet [19].

Free-form Text. includes human written knowledge, Wikipedia, and page text from search engines like Google, and WikiData. WikiData is entity-centric and each entity is associated with a description. Usually, human written knowledge is designed for one particular domain and thus might not apply to other domains. The other sources cover different domain knowledge and since they are from the internet therefore can be updated with time. Such knowledge might need to be pre-processed such as filtering tables, images, and some dirty words. In addition, such unstructured language might introduce noise or ambiguity compared to knowledge graphs [15].

Knowledge generation from Pre-Trained Models. In the realm of information retrieval, a significant shift is occurring from traditional explicit knowledge sources to the use of generated knowledge from large language models (LLMs), marking a move towards implicit knowledge. LLMs like GPT-3 and GPT-4, trained on diverse and extensive datasets encompassing a wide range of internet texts, books, and scientific papers, have brought about this paradigm shift. Unlike explicit knowledge, which is directly sourced and structured, LLMs represent a vast, albeit sometimes less precise, reservoir of knowledge stored across billions of parameters. However, this advancement comes with challenges such as lack of interpretability, where understanding the models' decision-making process is complex, and the risk of generating inaccurate information, known as "hallucinations". Despite these drawbacks, LLMs are increasingly used in various applications, including chatbots, content creation, and data analysis, although they are constrained by the quality and scope of their training data. As we continue to explore LLMs' capabilities and limitations in later chapters, we'll delve deeper into their role in generative information retrieval (GIR) and the ethical considerations surrounding their use.

3.2.3 Indexing

Indexing in information retrieval is the process of organizing data in a way that enables efficient access to information. There are various ways to index, including single-term, multi-term, inverted indexing, and semantic, each catering to different aspects of data organization. The indexing process involves meticulous document processing-tokenization, stemming, and stop-word removal-and the construction of an index based on term frequencies and document frequencies. Challenges in this field arise from managing large data volumes, dynamic content, and linguistic variations. Real-world applications, such as search engines and digital libraries, demonstrate the critical impact of indexing on user experience and information accessibility.

3.2.4 Scoring Function

The scoring function is a fundamental element that determines the relevance of documents to a user's query. Central to this function is not only the evaluation of keyword matches but also the assessment of semantic relationships between the query and target contents. Modern scoring functions consider factors like term frequency and inverse document frequency (TF-IDF) and extend to more nuanced measures that capture the semantic essence of texts. Techniques such as vector space models utilize representations like word embeddings, where the relevance is often quantified using measures like the inner-dot product or cosine similarity. These methods allow the scoring function to transcend beyond mere keyword alignment, enabling it to grasp the underlying contextual and conceptual similarities. This evolution in scoring functions, embracing semantic analysis, is pivotal for enhancing the accuracy and relevance of search results, especially in complex queries where contextual understanding is key to retrieving the most pertinent information.

3.3 Text Retrieval

Text retrieval remains the primary method for searching [20]. Moreover, text plays a crucial role in multimodal information retrieval; understanding the principles of text retrieval provides a foundation for the multimodal information retrieval topics that will be explored in subsequent sections. In the realm of text retrieval, we explore various representation techniques ranging from traditional bag-of-words models to advanced neural network-based embeddings. These representations are critical in capturing the semantic essence of the text, which is then leveraged by retrieval models such as TF-IDF [21], BM25 [22], and more recent deep learning approaches like Transformers-based models [23], such as BERT [24], which have revolutionized the accuracy and contextuality of text retrieval. Later on, we will discuss three types of fundamental text retrieval systems.

3.3.1 Text Representation

As the foundation for multimodal retrieval systems, effective text representation plays a vital role in capturing semantic information and establishing meaningful connections with other modalities. In this section, we will explore various techniques and approaches used for text representation, ranging from traditional methods to more advanced deep learning-based models.

We begin by discussing the early, yet essential, techniques for text representation, such as the Bag-of-Words (BoW) model and Term Frequency-Inverse Document Frequency (TF-IDF) weighting [21], which laid the groundwork for understanding and representing textual information. Next, we explore more advanced and sophisticated methods, including vector space models like Word2Vec [25] and GloVe [26]. These techniques leverage word co-occurrence and neural network architectures to generate dense word embeddings, enabling a richer representation of semantic relationships between words. Subsequently, we will discuss the role of recurrent neural network (RNN) models, such as Long Short-Term Memory (LSTM) [27] and Gated Recurrent Units (GRU) [28], in the context of text representation. These neural architectures, specifically designed to handle sequential data, are capable of capturing word order and long-range dependencies within text, addressing some limitations of earlier embedding techniques. Lastly, we will describe the state-of-the-art methods based on transformer [23] architecture and transfer learning, such as BERT [29], GPT [30]. These transformer-based models have revolutionized text representation by effectively capturing contextual information, syntactic structure, and semantic relationships, leading to significant advancements in various natural language processing tasks.

3.3.1.1 Bag-of-Words (BoW)

The process of creating a Bag-of-Words (BoW) representation involves several key steps to transform the raw text into a structured, fixed-length vector. We will describe each step with a walk-through example. Assuming that we have two simple documents: D1 = "The quick brown fox jumps over the lazy dog."; D2 = "Two elephants were discussing their favorite type of music."

1. Text preprocessing: Before creating the BoW representation, it is essential to preprocess the raw text to ensure a clean and consistent format. This typically involves:

    a. Lowercasing: Convert all characters in the text to lowercase to ensure uniformity and eliminate case sensitivity. Example, D1: "The quick brown fox jumps over the lazy dog." D2: "two elephants were discussing their favorite type of music."
    b. Removing special characters and punctuation: Remove any special characters, numbers, and punctuation marks from the text to focus on the words themselves. D1: "The quick brown fox jumps over the lazy dog" D2: "Two elephants were discussing their favorite type of music".
    c. Removing stop words: Eliminate common words that do not contribute much to the overall meaning of the text. These words, known as stop words, tend to be frequent and can create noise in the representation. To be an example, let's assume that stop words include "the", "over", "were", "their", and "of", and after removing the stop words, D1 becomes "quick brown fox jumps lazy dog", and D2 becomes "two elephants discussing favorite type music".

2. Tokenization: Break the preprocessed text into individual words or tokens. Tokenization can be done using whitespace, punctuation, or custom delimiters as separators. For example: D1 would be ["quick", "brown", "fox", "jumps", "lazy", "dog"] and D2 would be ["two", "elephant", "discussing", "favorite", "type", "music"].
3. Stemming or Lemmatization (optional): Reduce words to their root form to combine different forms of the same word and decrease the size of the vocabulary. Stemming involves removing inflections and affixes from words, while lemmatization reduces words to their base form according to the language's morphological rules. For example, D1 would be ["quick", "brown", "fox", "jump", "lazy", "dog"] and D2 would be ["two", "elephant", "discuss", "favorite", "type", "music"].

3.3 Text Retrieval

4. Vocabulary construction: Create a vocabulary by extracting all unique words from the preprocessed corpus. This vocabulary serves as a reference for encoding the documents. You can also limit the vocabulary size by selecting the top N most frequent words, where N is a predefined number. Since in this example, we have a small vocabulary size, we will just set N to be the unique words in this example, which results in a vocabulary: ["quick", "brown", "fox", "jump", "lazy", "dog", "two", "elephant", "discuss", "favorite", "type", "music"].
5. Vector encoding: Represent each document as a fixed-length vector using the constructed vocabulary. Each element in the vector corresponds to a word from the vocabulary, and its value can be one of the following:

   a. Binary encoding: Assign a value of 1 if the word is present in the document, and 0 if it is absent. D1: [1, 1, 1, 1, 1, 1, 0, 0, 0, 0, 0, 0], D2: [0, 0, 0, 0, 0, 0, 1, 1, 1, 1, 1, 1].
   b. Term Frequency (TF) encoding: Assign a value equal to the number of times the word appears in the document. D1: [1, 1, 1, 1, 1, 1, 0, 0, 0, 0, 0, 0], D2: [0, 0, 0, 0, 0, 0, 1, 1, 1, 1, 1, 1].
   c. Term Frequency-Inverse Document Frequency (TF-IDF) encoding: Assign a value that takes into account not only the frequency of the word in the document but also its rarity across the entire corpus. This method helps to emphasize more informative words and downplay common words that appear in many documents. We will have a more detailed explanation in the next subsection when we discuss TF-IDF.

6. N-grams (optional): Instead of using single words as tokens, you can create n-grams, which are contiguous sequences of n words from the text. Incorporating n-grams into the BoW representation can help capture some local context and word order information, at the cost of increasing the size of the vocabulary and the complexity of the representation.

Once the BoW representation is created, it can be used for various information retrieval tasks, such as document classification, clustering, and similarity measurement. Keep in mind that while the BoW model is simple and easy to implement, it has several limitations, including its inability to capture word order, syntactic structure, and semantic relationships. More advanced text representation techniques, such as word embeddings and transformer-based models, have been developed to address these shortcomings.

3.3.1.2 Frequency-Inverse Document Frequency (TF-IDF)

TF-IDF is a popular weighting scheme in information retrieval and text mining that reflects the importance of a term within a document in the context of a larger corpus. It combines two measures: Term Frequency (TF) and Inverse Document Frequency (IDF).

Term Frequency (TF) This represents the frequency of a term in a document. The higher the frequency, the more important the term is within that specific document. The term frequency can be calculated as the count of this term in a document.

Inverse Document Frequency (IDF) This measures the significance of a term across the entire corpus. The higher the IDF value, the more unique and informative the term is. The inverse document frequency is calculated as follows:

$$\text{IDF}(t) = \log(N/df(t)), \tag{3.3}$$

where N is the total number of documents in the corpus, and $df(t)$ is the number of documents containing the term t.

TF-IDF Encoding The TF-IDF encoding is the product of the term frequency and the inverse document frequency for each term in a document. This measure helps to identify terms that are important within a specific document but not necessarily common across the entire corpus.

$$\text{TF-IDF}(t, d) = \text{TF}(t, d) \times \text{IDF}(t) \tag{3.4}$$

Now, let's consider the same example we have in the previous section. After preprocessing,

D_1 :["*quick*", "*brown*", "*fox*", "*jump*", "*lazy*", "*dog*"] and

D_2 :["*two*", "*elephant*", "*discuss*", "*favorite*", "*type*", "*music*"].

The vocabulary consists of 12 unique terms:

["*quick*", "*brown*", "*fox*", "*jump*", "*lazy*", "*dog*", "*two*", "*elephant*", "*discuss*", "*favorite*", "*type*", "*music*"].

Now, we calculate the TF and IDF values for these terms:

- TF. D_1 : [1, 1, 1, 1, 1, 1, 0, 0, 0, 0, 0, 0], D_2 : [0, 0, 0, 0, 0, 0, 1, 1, 1, 1, 1, 1].
- IDF for each term. [0.301, 0.301, 0.301, 0.301, 0.301, 0.301, 0.301, 0.301, 0.301, 0.301, 0.301, 0.301].
- TF-IDF Encoding. D_1 : [0.301, 0.301, 0.301, 0.301, 0.301, 0.301, 0, 0, 0, 0, 0, 0], D_2 : [0, 0, 0, 0, 0, 0, 0.301, 0.301, 0.301, 0.301, 0.301, 0.301].

3.3.1.3 Word2Vec

Word2Vec is used for converting words into embeddings, which are dense numerical representations of words in a continuous vector space. Such vectors are designed to capture

3.3 Text Retrieval

semantic and syntactic similarities between words. Simply, Word2Vec is a neural network with two-layer perceptrons. There are two main approaches to training Word2Vec: the Continuous Bag of Words (CBOW) model and the Skip-gram model. Both CBOW and Skip-gram models are trained using a large corpus of text data.

Continuous Bag of Words (CBoW). In CBoW [25], the model predicts a target word based on its context, which consists of the surrounding words. The input to the model is a window of context words, and the output is the target word. The context words are transformed into their corresponding word vectors. These word vectors are averaged to obtain a single vector representation, which is then used to predict the target word. The model's objective is to maximize the probability of predicting the target word correctly given the context.

Skip-gram. The Skip-gram model [25], on the other hand, aims to predict the context words based on a target word. Given a target word, the model tries to predict the words that are likely to appear in its context. The target word is transformed into its word vector, which is then used to predict the context words. The objective of the Skip-gram model is to maximize the probability of predicting the context words correctly given the target word.

3.3.1.4 GloVe

Similar to Word2Vec, GloVe [26] also aims to capture the semantic and syntactic relationships between words by representing them as vectors in a continuous vector space. The distinction from Word2Vec is in its approach to generating word embeddings: while Word2Vec is based on a neural network architecture, GloVe is a count-based model that leverages global word co-occurrence statistics. The training can be broken down into the following four steps.

- Co-occurrence Matrix: GloVe starts by constructing a co-occurrence matrix from a large corpus of text. The matrix captures the frequency of word co-occurrences across the entire corpus. Each entry in the matrix represents the number of times two words co-occur within a predefined context window.
- Probability Ratios: GloVe then transforms the co-occurrence counts into probabilities. It calculates the ratio of co-occurrence probabilities for a pair of words, considering both their co-occurrence frequency and the overall statistics of their individual word occurrences.
- Objective Function: GloVe defines an objective function that measures the difference between the dot product of word vectors and the logarithm of the co-occurrence probabilities obtained in the previous step. The goal is to minimize this difference, effectively optimizing the word vectors to capture the desired relationships between words.
- Training: The objective function is optimized using iterative training methods, such as stochastic gradient descent (SGD), to update the word vectors. The training process adjusts the vectors to minimize the objective function across the entire co-occurrence matrix.

3.3.1.5 Long Short Term Memory Network (LSTM)

LSTM [27] is a type of recurrent neural network (RNN) architecture that is designed to handle the vanishing gradient problem and capture long-term dependencies in sequential data. LSTMs were introduced to overcome the limitations of traditional RNNs, which struggle with preserving and propagating information over long sequences. In tasks involving text, speech, or time series data, LSTMs are particularly effective. LSTMs have proven to be effective in a variety of sequential data tasks, such as natural language processing (NLP), speech recognition, machine translation, and time series analysis. They excel in capturing dependencies that extend over long time lags, making them suitable for tasks that require modeling complex temporal relationships. There are five important components in LSTM:

- Cell State (C): The cell state serves as the information highway of an LSTM. It runs linearly through the entire sequence, with minimal alterations, allowing information to flow without much degradation. The cell state can be considered as the long-term memory of the network.
- Gates: LSTMs utilize various gating mechanisms to control the flow of information and regulate the cell state. These gates include:
- Forget Gate: Decides which information from the previous cell state should be discarded or forgotten. Input Gate: Determines which new information should be stored in the cell state.
- Output Gate: Filters the information from the current cell state to produce the output of the LSTM.
- Hidden State (h): The hidden state, also known as the output state, carries the relevant information from the current input and previous hidden state. It acts as the short-term memory of the network and is influenced by the input, previous hidden state, and the current cell state.

The LSTM architecture incorporates these components to process sequential data. At each time step, the LSTM takes an input, the current time step's input (x), and the previous hidden state (h), and calculates new values for the cell state (C) and the hidden state (h). These calculations involve element-wise operations and activation functions. During training, the parameters of the LSTM, including the weights and biases, are learned through backpropagation and gradient descent methods, optimizing the network's performance for a specific task.

3.3.1.6 Transformers

Transformers [32] have revolutionized the field of natural language processing (NLP). It heavily relies on self-attention. Transformers are composed of a stack of identical layers. Each layer contains two sub-layers: a multi-head self-attention mechanism and a position-wise feed-forward neural network. The self-attention mechanism allows the model to weigh the importance of different words in a sequence when encoding or decoding. It computes attention scores between each word and all other words in the sequence, capturing dependencies and relationships. This is achieved by transforming the input sequence into three parts: queries, keys, and values. Then, the attention scores are computed as the dot products between the queries and keys, scaled, and passed through a softmax function to obtain the attention weights. Finally, the values are weighted by the attention weights and summed to obtain the output.

The multi-head attention mechanism performs the self-attention operation multiple times in parallel, each with different learned linear projections of the queries, keys, and values. This allows the model to capture different aspects of the input representations.

The position-wise feed-forward neural network consists of two linear transformations with a non-linear activation function in between. It is applied to each position independently, which allows the model to capture interactions between different dimensions of the representation.

The original transformer has an encoder and decoder, where the decoder is slightly different from the encoder. The encoder processes the input sequence, while the decoder generates the output sequence autoregressively, attending to the encoder's output at each step.

Since the Transformer came out, varients language models have been proposed based on the Transformer architecture. BERT is the first language model with novel pretraining strategies that allow it to learn powerful contextualized word representations. It employs two pre-training tasks: masked language modeling (MLM) and next sentence prediction (NSP).

- In MLM, BERT randomly masks a certain percentage of the input words and learns to predict them based on the context of the surrounding words. This helps BERT to acquire a deep understanding of word relationships and capture fine-grained contextual information.
- In NSP, BERT takes pairs of sentences as input and learns to predict whether the second sentence is the actual subsequent sentence in the original document or a random sentence from the dataset. This task helps BERT to learn relationships between sentences and to grasp broader contextual dependencies.

BERT includes a pooling mechanism that aggregates the contextualized representations of individual words to obtain a fixed-dimensional representation for the entire input sequence. BERT employs the [CLS] token to represent the entire input text sequence, which can be used for global representation and classification tasks.

Following the introduction of BERT, a multitude of variations have emerged in the field of natural language processing (NLP), showcasing diversity in tokenization methods. While BERT employs WordPiece tokenization, other approaches like Byte-Pair Encoding (BPE), SentencePiece, Word tokenization, and Character tokenization have been proposed. Additionally, various attention mechanisms have been explored. BERT utilizes global attention, where each word attends to every other word. On the other hand, alternative models incorporate local attention, restricting word attention to specific regions within the input sequence rather than considering all positions simultaneously. Furthermore, these models differ in terms of training corpus and input length.

3.3.2 Sparse Retriever

Sparse retrieval methods produce vectors having a relatively small number of non-zero elements hence the name sparse. One of the most widely known algorithms for sparse retrieval is **Best Match 25 (BM25)** [22]. It is a family of scoring functions that ranks a set of documents based on the query terms appearing in each document. Specifically, the calculation of BM25 involves Term Frequency with Saturation Function, Inverse Document Frequency, and Document Length Normalization:

Term Frequency (TF): TF corresponds to the number of times a particular term appears in a document. BM25 uses a modified term frequency that takes into account saturation effects to prevent overemphasis on the heavily repeated terms as very high frequencies often correspond to less informative terms.

Inverse Document Frequency (IDF): IDF estimates the importance of a term in the entire corpus by assigning higher weights to those terms that are rare in the corpus and lower weights to those that are common. IDF for a term t is defined as $\log((N - n + 0.5)/(n + 0.5))$, where N is the total number of documents and n is the number of documents containing the term t.

Document Length Normalization: BM25 incorporates document length normalization to address the impact of document length on similarity scoring. Longer documents tend to have more occurrences of a term, which may lead to a potential bias. Document length normalization counters this bias by dividing the term frequency by the document's length and applying a normalization factor.

One of the popular variations of the BM25 score for a document D concerning a query Q is calculated as:

$$\text{BM25}_{\text{score}}(D, Q) = \sum_{i=1}^{N} IDF(q_i) \cdot \frac{f(q_i, D) \cdot (k_1 + 1)}{f(q_i, D) + k_1 \cdot (1 - b + b \cdot \frac{|D|}{\text{avg } dl})}$$

where, IDF(q) represents the inverse document frequency of the query term q, TF(q, D) denotes the modified term frequency of the term q in document D, |D| represents the length of document D, and avg dl is the average document length in the corpus. Parameters k_1 and b are constants that control the impact of term frequency saturation and document length normalization, respectively.

3.3 Text Retrieval

Advantages of BM25

- **Simplicity**: BM25 is a widely used ranking algorithm due to its simplicity and effectiveness in producing relevant search results. It also takes into account both term frequency and document length normalization, which helps address the issue of document length bias.
- **Scalability**: BM25 can handle large document collections efficiently, making it a scalable technique for real-world search applications.
- **Indexing and Retrieval Speed**: BM25 is usually much faster than modern deep learning-based approaches for both indexing (calculating document vectors) and retrieval.
- **Explainability**: The sparse vectors are easily interpretable and allow checking why a particular document was retrieved for a given query and what terms had the highest contribution. It is even possible to influence the results by simply deleting or updating particular terms or adding custom rules to calculate the term weights.

Limitations of BM25

- **Lack of Semantic Understanding**: BM25 does not consider the semantic meaning or context of the query and the documents, which may result in sub-optimal document rankings.
- **Term independence Assumption**: BM25 assumes statistical independence between query terms, which may not hold in some cases where term dependencies exist.
- **Reliance on Term Frequency and Document Length**: BM25 heavily relies on term frequency and document length, thus ignoring other important factors like document structure and relevance feedback.

3.3.3 Dense Retriever

Sparse retrieval methods have some advantages like simplicity and explainability; however, the sparse representation of text based on tokens fails to represent its complete semantics. Thus, the sparse methods which are usually based on term occurrences may fail to map vectors corresponding to even synonyms or paraphrases of a text that consist of completely different tokens closely in the representation space. In contrast, the dense representation methods have been shown to encode the semantic meaning of the text to a great extent and can thus result in better retrieval. Dense encodings are also learnable by adjusting the embedding functions, which provides additional flexibility to have a task-specific representation.

One of the earliest works that adopts large transformer-based models for dense retrieval is **Dense Passage Retriever (DPR)** [3]. DPR uses a dense encoder $E_P(\cdot)$ which maps a text passage to a d-dimensional real-valued vector and builds an index for all the passages in the corpus. At run-time, DPR applies a different encoder $E_Q(\cdot)$ that maps the input query to a d-dimensional vector and retrieves the passages of which vectors are the closest to the query vector. The similarity between the query and the passage is simply the dot product of their vectors:

$$\text{sim}(q, p) = E_Q(q)^T E_P(p).$$

Training the Encoders: The objective is to create a vector space such that relevant pairs of queries and passages will have smaller distances (i.e., higher similarity) than the irrelevant ones, by learning a better embedding function. In DPR, two independent BERT models are used as encoders for queries and passages. The representation of the '[CLS]' token of the output layer of BERT is used as the encoding of the corresponding text. The whole bi-encoder model is trained by optimizing the loss function as the negative log-likelihood of the positive passage.

In DPR, a dataset of question-and-answer pairs is used for training. Usually, negative examples are not available in a dataset and need to be sampled. As negative examples, DPR uses passages from the same mini-batch and one passage returned by BM25 which don't contain the answer but match most question tokens (in-batch negatives). This strategy of mining negative examples can make the computation efficient while achieving great retrieval performance. It is an effective strategy for learning a dual-encoder model that boosts the number of training examples.

Let D be the training dataset that consists of m instances:

$$D = \{< q_i, p_i^+, p_{i,1}^-, \ldots, p_{i,n}^- >\}_{i=1}^m. \tag{3.5}$$

Each instance contains one query q_i and one relevant (positive) passage p_i^+, along with n irrelevant (negative) passages $p_{i,j}^-$. The negative log-likelihood of the positive passage is used as the loss function.

$$L(q_i, p_i^+, p_{i,1}^-, \ldots, p_{i,n}^-) = -\log \frac{e^{\text{sim}(q_i, p_i^+)}}{e^{\text{sim}(q_i, p_i^+)} + \sum_{j=1}^n e^{\text{sim}(q_i, p_{i,j}^-)}} \tag{3.6}$$

One of the limitations of such a retrieval method is that it incurs a large memory cost due to the massive size of its passage index, which must be stored entirely in memory at runtime. Addressing this problem, [31] proposed Binary Passage Retriever (BPR), which learns to hash continuous vectors into compact binary codes using a multitask objective that simultaneously trains the encoders and hash functions in an end-to-end manner. Specifically, BPR integrates a learning-to-hash technique into the DPR architecture to reduce the size of the passage index by storing it as binary codes.

3.3.4 Hybrid Retriever

Sparse and dense retrievers each bring their unique advantages and limitations. However, a hybrid approach that combines these two methodologies tends to outshine each used separately by harnessing their collective strengths. This combined effectiveness has popularized hybrid retrievers in various real-world scenarios, as evidenced by studies such as [33, 34]. Hybrid Retrievers integrate sparse and dense retrieval tactics, blending the relevance scores from both for document ranking. Typically, this hybrid ranking is achieved through a linear amalgamation of scores from both sparse and dense retrievers, as outlined in [3, 35–38].

Moving beyond straightforward score combination, Chen et al. [39] adopt Reciprocal Rank Fusion [40] to determine final rankings, focusing on each candidate's position as retrieved by individual retrievers. Additionally, Arabzadeh et al. [41] have engineered a classification model to ascertain the most effective retrieval strategy, whether it be sparse, dense, or a hybrid of the two.

Beyond mere score amalgamation, some studies also employ sparse retriever cues to refine dense retriever training, as noted in [34, 42]. Yet, many hybrid models are dependent on dense retrievers which demand significant indexing memory. An innovative departure from this trend is the method by Ma et al. [37], incorporating strategies like linear projection, PCA, and product quantization [43] to lessen the memory demands of the dense retriever.

Luo et al. [44] further mitigated the indexing load of cumbersome retrievers through knowledge distillation and multi-task learning, coupling this with a sparse retriever such as BM25. Another advantage of is to achieve generalization [45]. The primary focus of the following discussion is on hybrid methods that fuse two scores, representing a simple yet potent approach as presented in [44].

Reranking by Sparse and Dense Retriever Scores. In this approach, a combined model reevaluates and selects the final documents by amalgamating the scores from both a dense and a sparse retriever. The process begins with the sparse retriever, like BM25, which retrieves the top-k documents. This is followed by a similar step using a dense retriever, such as DPR. During the third stage, each document receives a score from both retrievers, denoted as S_{BM25} and S_{DR}, respectively. To standardize these scores, a method like MinMax normalization is applied, resulting in normalized scores S'_{BM25} and S'_{DR}, which range between [0, 1]. In the last step, each document is assigned a revised score for final ranking:

$$S_{\text{hybrid}} = w_1 \times S'_{\text{BM25}} + w_2 \times S'_{\text{DR}}, \tag{3.7}$$

Here, w_1 and w_2 are the weights assigned to the BM25 and DR scores, respectively. These weights can be determined either by fine-tuning them on specific downstream tasks or by assigning them equal values. If a context is not retrieved by one of the retrievers, its score from that retriever is set to 0.

3.4 Multimodal Retrieval

As we venture into the realm of multimodal information retrieval, it becomes crucial to establish a solid foundation for representing, processing, and integrating diverse modalities in a unified retrieval framework. This section aims to provide a comprehensive understanding of the various models and methods that have been developed to achieve this goal. We'll begin with an overview of advanced transformer-based multimodal systems, followed by illustrations of representative systems designed for multimodal query retrieval tasks.

3.4.1 Advanced Multimodal Models and Pretraining Tasks

One important direction to improve the performance of such models on retrieval tasks is by pretraining the models with more advanced techniques so that the text and image encoder can align better. In the following, we introduce multiple advanced vision and language models that go beyond the basic models described in Sect. 3.2.

3.4.1.1 Bootstrapping Language-Image Pre-training for Unified Vision-Language Understanding and Generation

BLIP [46] is a new Vision-Language Pre-training (VLP) framework designed to be flexible in transferring to both vision-language understanding and generation tasks and aims to excel in both. One of the most significant contributions of BLIP is the data bootstrapping techniques to filter noisy data to train the model.

Architecture. BLIP has an image encoder, a text encoder, an image-grounded text encoder, and an image-grounded text decoder. The image encoder and text encoder are based on existing models, ViT [47] and BERT [24] separately. In the image-grounded text encoder, to inject visual information with the text input, one additional cross-attention (CA) layer is inserted between the self-attention (SA) layer and the feed-forward network (FFN) for each transformer block of the text encoder. The image-grounded text decoder simply replaces the bi-directional self-attention layers in the image-grounded text encoder with causal self-attention layers. The image-grounded encoder and decoder share the same parameters except for the SA layers.

Training Strategies and Data. BLIP is pretrained on the existing image-pair datasets with three tasks, image-text contrastive learning, image-text matching, and image-conditioned language modeling. Not only that, BLIP incorporates a bootstrapping technique to create large-scale data which is then used to further train BLIP (it is known that large-scale training data is helpful.) Specifically, bootstrapping involves two processes, an image captioning generation process, and a filtering process. First of all, they pre-train BLIP on COCO datasets,

3.4 Multimodal Retrieval

then the pretrained image-grounded text decoder in BLIP is taken as a caption generator that generates a caption for any image from the Web. And the pretrained image-grounded text encoder is used to filter both the original web texts and the synthetic caption. Through these two steps, they generate a new dataset called CapFilt.

Downstream Tasks. BLIP is evaluated on both text-image retrieval tasks (COCO and Flickr30K), image captioning tasks (COCO and No- Caps), VQA, Natural Language Visual Reasoning, and Vision Dialogue (VisDial). Among all the tasks, BLIP achieves better results than the SOTA model by the model published. Additionally, they conduct a zero-shot evaluation for text-to-video retrieval and video question-answering tasks. Despite not encoding temporal information and not being trained on the specific downstream domain, BLIP demonstrates robust zero-shot generalization.

Conclusion. As demonstrated by their experiments, BLIP shows the effectiveness of their data generation and filtering strategy, which achieves new state-of-the-art results on the retrieval and reasoning tasks in the multimodal domain. The authors also highlight a few future directions: multiple rounds of bootstrapping; generating multiple images to enlarge the training data; and model ensemble.

3.4.1.2 GLIP

GLIP [2], designed for grounded language-image pretraining, integrates objectives from both object detection and phrase grounding.

Architecture. GLIP incorporates both an image encoder and a text encoder. The image encoder utilizes DyHead [48], an object extraction model that identifies object feature regions or boxes. This feature is then channeled to two prediction heads: one for class label determination through a classification layer and the other for pinpointing the bounding box location using a regression layer. The language encoder is based on a BERT model [29]. Moreover, to enhance phrase grounding performance and prompt-aware detection, GLIP employs a language-aware deep fusion approach. This introduces a cross-modality multi-head attention module (X-MHA) preceding both the image and text encoder layers. The X-MHA layer takes input as the representation of the object feature and the token feature from the last layer, and produces a cross-modality image feature and token feature by attending to the other modality feature. Such cross-modality features are subsequently combined with single-modality features, which are then directed to the individual modality encoder. Due to this deep-fused encoder, GLIP outperforms CLIP in retrieval tasks. However, this advantage is accompanied by a reduction in efficiency.

Training Strategies and Data. GLIP is pretrained on both object detection and phrase grounding. This unified pretraining approach allows GLIP to leverage more data.

To facilitate this, GLIP transforms object detection tasks into phrase grounding tasks. Specifically, the text encoder receives a textual prompt composed of combined class name candidate phrases to be grounded. The alignment score, derived from the dot product of the visual region and language tokens features, for the pertinent bounding box and class tokens, is optimized, while the scores for other class tokens are reduced. GLIP initially undergoes pretraining using the Objects365 dataset [49] for detection, GoldG [50] for grounding, and CC [51] and SBU [52] for caption. Subsequently, a GLIP model, once pretrained, acts as a 'teacher' model. This teacher model is employed to create bounding boxes for web-sourced image-text data. The generated data, combined with the initial human-annotated datasets, are then used to train a 'student' model. This student model leverages both the original human-annotated data and the additional data generated by the teacher model, enhancing its learning and performance.

Downstream Tasks. GLIP is evaluated on various tasks including object detection in datasets like Ms-COCO and LVIS, as well as phrase grounding in Flickr30K. Furthermore, various GLIP models were trained on a different combination of the pretraining data. In all these tasks, GLIP consistently demonstrates top-notch performance, both in zero-shot scenarios and when fine-tuned. The experimental results notably highlight the substantial benefits derived from incorporating grounding tasks in its pretraining phase.

Conclusion. GLIP transforms the task of object detection into a grounding challenge. Unlike standard object detection, grounding data encompasses a broader range of visual concepts. This wider scope enables GLIP to identify and detect more rare classes.

3.4.1.3 Flamingo: Bridge Language and Vision Model to Enable Few-Shot Learning

Flamingo [53] is a Visual Language Model (VLM) designed for adaptation to few-shot learning. Flamingo merges pretrained vision-only and language-only models, manages mixed sequences of visual and textual data, and accepts both images and videos. The adaptability of Flamingo allows it to be trained on large multimodal web datasets, enhancing the few-shot learning capabilities.

Architecture. Flamingo processes text interleaved with images or videos and returns text as its output. It incorporates a vision encoder, a Perceiver sampler [54], and a language model. The vision encoder, based on ResNET [55], generates a 2D grid of features. This grid is then converted into a 1D sequence. The Perceiver sampler, which is built on a Transformer architecture, produces a fixed number of image tokens from any vision feature given by the encoder. To achieve this, the Perceiver learns a set number of latent query features and cross-attends to the flattened visual elements. Following this, a pre-trained language model interprets both the text and the visual tokens, producing a text-based result. Within

3.4 Multimodal Retrieval

the language model's transformer structure, a newly trained gated cross-attention block has been inserted. In this layer, the visual tokens supply the key and value, while the language input provides the queries. Before adding the model's output with the textual tokens through the residual connection, a $tanh(\alpha)$ activation function is employed on the cross-attention layer's output to ensure the language model is intact. This language model is based on the Chinchilla models [56] and comes in varying sizes.

Training Strategies and Data. Flamingo is pre-trained a mixture of three kinds of datasets, all scraped from the web: an interleaved image and text dataset derived from webpages (MultiModal MassiveWeb, M3W dataset), image-text pairs, and video-text pairs. The M3W is collected from 43 million of webpages, contributing to the model's few-shot learning ability. For the image-text pairs, they first utilize the ALIGN dataset, and in addition, they collect a long-text image dataset (LTIP) with 312 million image-text pairs. Similarly, for the video-text pairs, they collected 27 million short videos (22 s) paired with text.

Conclusion. Flamingo set new state-of-the-art results on few-shot learning on a wide range of open-ended vision and language tasks and even better than fine-tuned models on some tasks.

3.4.1.4 PaLM-E

PaLM-E [57] is an embodied multimodal language model proposed to solve complex real-world challenges such as robotic problems where groundings are important. PaLM-E integrates the real-time sensor modality into the language model such that the input is a multimodal sentence that interleaves visual, continuous state estimation, and textual and the output is a natural language that solves the input task.

Model Architecture. The major components of PaLM-E are the PaLM-540B language model and the ViT-22B vision model. PaLM-E uses ViT to transform an image into a sequence of token embeddings, and they introduce multiple encoders ϕ (each encoder can be used for different positions) to project such image embeddings to the language embeddings. In addition, PaLM-E can also consume the state input such as color and size. Such input is simply represented by a vector and an encoder is applied to project this vector in the language space. During the training time, PaLM-E freezes the large language model and only trains the projection encoder ϕ.

Training Strategies and Data. PaLM-E is pretrained on a mixture of general vision and robotic tasks. The V&L tasks include VQA, GQA, and COCO. Webli, VQ2A, VQG, CC3M, ObjectAwar, OKVQA, VQAv2, COCO and Wikipediatext. It is also trained on 10% of the robotic tasks.

Downstream Task Performance. PaLM-E is evaluated on robotic planning tasks (e.g. Say-Can), general V&L tasks such as VQA (Ok-VQA) and image captioning, and the general natural language understanding and generation tasks (NLG/NLU). On the multimodal tasks, the model performance improved compared to the state-of-the-art models, however, on the NLG and NLU tasks, the performance decreased.

Conclusion. PaLM-E extends a Language and Vision Model (LLM) into an Embodied Multimodal Language Model, showcasing remarkable performance on intricate robotic planning challenges and Visual and Language (V&L) tasks. PaLM-E further demonstrates impressive capabilities in zero-shot, few-shot, and multimodal chain-of-thoughts generalization, while also excelling in generalization across multiple images when trained on a single image input dataset. However, it's worth noting that the underlying LLM faces a significant challenge known as catastrophic forgetting.

3.4.2 Mutlimodal Query for Image Retrieval

In this particular task, the search involves a multimodal query where the target outputs are images. A key application of this multimodal query in image retrieval is to search for an image which resembles the source image but with distinct features, as described in the accompanying text. For instance, the target image might feature a white dress instead of a blue dress, furthermore, the target image should not change other features in the source image. To achieve this, ComposeAE [58] introduces a two-step approach. Firstly, they employ a BERT encoder to extract textual features and a CNN model (e.g., Resnet) to obtain image features. Subsequently, they utilize a text-image compositional modular technique to derive the rotational representation of the image. This process generates an element-wise rotation of the source image in another complex space, yielding the resulting image representation. This resultant image feature is then utilized to identify and locate the target image.

3.4.3 Mutlimodal Query for Text Retrieval

The prior section covered image retrieval through multimodal queries. This section shifts the focus to retrieving text using multimodal queries. Initially, an overview of the three primary themes in multimodal query retrieval will be presented. Subsequently, the methods, experiments, and conclusions from the study [59] will be examined in detail in Sect. 3.4.4. Similarly, the proposals from [11] will be discussed in Sect. 3.4.5, providing a comprehensive analysis of these approaches.

Only use the text question. In the case of a multimodal query, the text component is typically utilized to retrieve the most relevant documents using a text-based retriever, such as

3.4 Multimodal Retrieval

BM25 or popular search engines like Google Search, which offer a vast amount of general knowledge. While being simple, it completely ignores the image information, thus, the retrieved document might be insufficient.

Convert the Image to the Captioning. To extract information from an image, a captioning generator is utilized to convert the visual content into a textual description. Pretrained captioning models like Oscar are commonly employed for this task. Once the caption is generated, it can be combined with the provided textual question and used as input for a text-based retriever. This approach leverages the comprehensive knowledge exhibited by large language models. Recent studies have demonstrated that when given a question, models like GPT-3 can generate substantial knowledge, showcasing their ability to provide informative responses.

Crossmodality Model. In recent studies, researchers have leveraged multimodality models to encode both text and images and produce representations that capture information from both modalities. This approach offers advantages over previous methods by preserving more image-related details. Different types of crossmodality models have been explored. Some utilize feature extraction models like Fast-RCNN to extract relevant features from the images. These extracted features are then combined with text representations to generate cross-modality representations. On the other hand, certain approaches directly incorporate image patches as input without relying on an additional feature extraction module. This allows for a more streamlined and efficient fusion of text and image information within the model itself.

3.4.3.1 Use Text-Image to Retrieve Explicit Textual Knowledge

While the input query is composed of text and image, the image can be converted to text format and then concatenated with the text query to form a new text query which can be fed to a text-based retriever to find the explicit textual knowledge. Luo et al. [59] use BM25 as the text retriever. On the other hand, Gao et al. [60], Luo et al. [59], Qu et al. [61] have applied DPR (described in Sect. 3.3.3) as the retriever, differently, TRIG [60] use the off-the-shelf retriever that is trained on the NQ dataset, and Luo et al. [59] have trained the DPR on their down-stream tasks.

3.4.3.2 Use Text-Image to Retrieve Implicit Textual Knowledge

Implicit knowledge refers to the information that is stored in model parameters. Since advanced transformer-based models have been trained on a large scale of pretraining data via a self-supervised manner and have a large number of model parameters, certain knowledge is stored in the model and can be generated with the proper prompt. For example, the earliest work which shows language model as a knowledge base in [62], where they show that given the input "Dante was born in [MASK]", BERT model can decode [MASK] as *Florence*.

Some works have taken advantage of large language models such as GPT-3 that implicitly store large amounts of knowledge. These language models only take the text as input and can not be locally fine-tuned on downstream tasks. Gui et al. [63], Yang et al. [64] takes the off-the-shelf GPT-3 to retrieve the knowledge. To do so, the first step is to convert the image to text. The text forms can be the image captioning, the image object names, and the text in the image. External models are needed to be utilized to get such information: captioning generation model, object detection models, and optical character recognition (OCR) models. After the image is converted to the textual information, they are concatenated with the textual question as the model input to the GPT-3, and then GPT-3 can generate the text, which is considered as the retrieved knowledge.

3.4.3.3 Use Image to Retrieve Explicit Textual Knowledge

In this section, we walk through the system KAT [63], serving as an exemplary model for retrieving information from multiple corpora through text and image retrieval separately for explicit and implicit information. From a mathematical perspective,

$$Knowledge = R_I(Image, C1) \oplus R_T(Text, C2), \tag{3.8}$$

where R_I is a retriever that takes $Image$ as input and retrieves knowledge from the corpus $C1$, and R_T is a retriever that takes $Text$ as input and retrieve knowledge from the corpus $C2$.

For the $Text$ part, KAT transforms the image into a corresponding caption which is then concatenated with the text question. This concatenated input is used to prompt GPT-3 to generate knowledge. Additionally, the input prompt includes demonstrations that act as in-context learning examples for GPT-3.

Image Patches to Text Gui et al. [63] divides the image into patches to find the most relevant knowledge from a corpus that is constructed from Wikidata (each entry is composed of the entity and the corresponding description). Particularly, Given an image I, KAT utilizes a sliding window with stride to generate N image patches represented by $\{v_1 \cdot v_j\}$. Such patches are fed to an image encoder (E_{img}) to get the corresponding patch-dense vector representation which is in d dimension. KAT utilizes the image encoder in CLIP [1]. Given any entry in the knowledge corpus, a text encoder (E_{text}) to obtain the corresponding dense representation, which is also in d dimension, so that the similarity score of any entry related to the image patch can be obtained by the inner-dot product of two dense representation as expressed in below. KAT uses the text encoder in CLIP and gets the representation of the special token $[CLS]$.

$$Score(v^i, T) = E_{img}(v^i)^\top E_{text}(T) \tag{3.9}$$

3.4 Multimodal Retrieval

KAT uses each patch to find k knowledge from the corpus and combine it all together which yields $N \times k$ knowledge in total. Another alternative way to do it is by using the global representation of the image to find k knowledge.

3.4.4 Visual and Language Dense Passage Representation

Luo et al. [59] designed three multimodal query retrievers to tackle open domain knowledge-based visual question-answering tasks (OkVQA [8]). The multimodal-query retrievers will retrieve corresponding knowledge for the multimodal queries, which can then be used in a question-answering model to generate the answers. In the following, we will give details of three multimodal retrievers.

3.4.4.1 Term-Based Retriever

Text-based retrieval systems, such as the BM25 model [65], are established concepts in information retrieval. BM25 treats queries and documents as sparse vectors in a d-dimensional space, where d represents the vocabulary size. It ranks query-document pairs based on term frequency. More details on the BM25 model are available in Sect. 3.3.2. However, BM25's limitation lies in its exclusive focus on text queries, which is a significant drawback in contexts like OKVQA, where queries also include images. Relying solely on BM25 neglects the informational content of images.

To address this gap, the authors initially employed a caption generation model to produce image captions. Specifically, they use the Oscar [66] model, which is pretrained on 6.5 million text-tag-image triples from various sources, including COCO [5], Conceptual Captions (CC) [51], SBU captions [52], flicker30k [67], and GQA [68]. These datasets provide images from domains similar to those in the OKVQA dataset. Subsequently, the authors merge the original question with the generated image caption to form an enhanced query. This composite query is then processed by the BM25 model to retrieve relevant documents (Fig. 3.1).

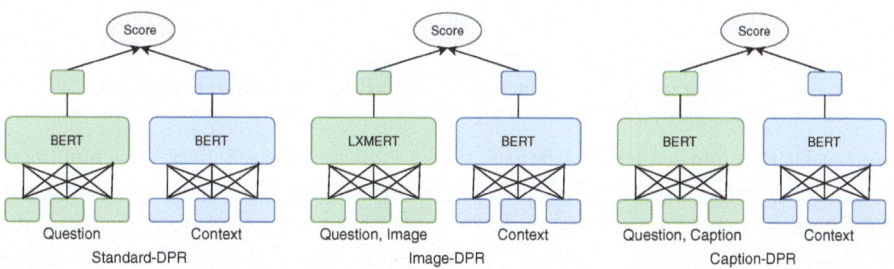

Fig. 3.1 This figure explains how BM25 handles multimodal queries with the assistance of a captioning generator. will change later

3.4.4.2 Visual Neural Retriever

In contrast to the BM25 model, neural retrievers like the Dense Retriever (DPR) [3] used in this research, leverage neural networks to generate dense representations of queries and contexts. DPR employs a pair of BERT [24] models to independently encode the query and context, using an inner-dot product to estimate their correlation. Similar to BM25, DPR processes queries in text format. For more information on the DPR model, see Sect. 3.3.3.

To adapt DPR for use in the visual domain, the authors suggest two approaches. Both methods share the same query and context encoders, utilizing the BERT language model for encoding knowledge. The subsequent section will detail the query encoder used in each of these methods.

Image-DPR incorporates LXMERT [69] as its multimodal-query encoder, which is composed of three distinct modules: a text encoder, an image encoder, and a cross-modality encoder. The text and image encoders focus on their respective single modalities, while the cross-modality encoder integrates these to form a unified representation. The text encoder utilizes the wordpiece tokenizer, breaking down sentences into word sequences identified by unique tokens. Each word is linked to an index indicating its position in the sentence. These word identifiers are transformed into embedding vectors, while the index is processed through a transformation function. The resulting word embedding and transformed index are merged and normalized to create text input embeddings. The core of these encoding layers is the Transformer architecture [23], essentially a series of transformer blocks comprising a self-attention layer followed by two Multi-Layer Perceptions (MLPs). For image encoding, each image is represented by a collection of objects identified by an object detector, providing both region-of-interest (ROI) features and bounding box coordinates. These features and coordinates are processed through two-layer MLPs, and their averages form the image input embeddings. The encoding layers here mirror the text encoder, consisting of stacks of transformer layers. The cross-modality encoder is built with blocks containing a cross-attention layer, a self-attention layer, and an MLP. The cross-attention layer utilizes the output of the last block's features from both language and vision streams, facilitating the exchange and alignment of information across modalities for learning joint cross-modality representations. LXMERT outputs for language, vision, and cross-modality are then generated. A special [CLS] token precedes the sentence words, and the feature vector corresponding to this token in the language feature sequence is used as the cross-modality output.

Caption-DPR employs a BERT language model as its text encoder to process multimodal queries. In a manner akin to Term-based retrieval (Sect. 3.4.4.1), this method utilizes the OSCAR caption generation model to derive captions for images. These captions are then integrated with the question, forming a new, combined query. Following this, a standard BERT model is applied as the query encoder, tasked with generating the query's representation.

3.4 Multimodal Retrieval

Weak-supervised Training Strategies In this approach, the authors evaluate their retriever on the Ok-VQA benchmark, which provides only the answers to questions but not the specific knowledge needed for those answers. Consequently, the retriever is trained under weak supervision due to the absence of known ground-truth knowledge context for each question-image pair. Specifically, when presented with a query and an image, the authors operate under the assumption that any knowledge containing the answer is relevant. For training purposes, they employ in-batch negative samples [3]. This means that during training, any relevant knowledge for other questions within the same batch is treated as irrelevant. The training of the model is executed using contrastive learning. They utilize the inner-dot product function to evaluate the similarity score between relevant and irrelevant knowledge pertaining to a question. The model is then optimized to maximize the score for question-positive knowledge pairings, employing negative log-likelihood for this purpose.

3.4.4.3 Experiments and Conclusion

They applied the proposed retrievers to find the external knowledge and evaluate the OkVQA dataset. The experiments show that among different variations of the proposed retriever, Caption-DPR achieved the best performance based on Precision and Recall. Caption-DPR consistently outperforms BM25 and Caption-DPR on the various number of retrieved knowledge. It is interesting to see that Caption-DPR outperforms BM25 significantly since BM25 is a hard-to-beat baseline in open-domain QA [70]. It indicates that neural retriever has better application than term-based retrieval methods in the vision domain (Table 3.1).

3.4.5 End-to-End Multimodal Queries Retrieval

Presently, multimodal-query Retrievers, referenced in [59, 61, 71], commonly adopt a dual-step process for retrieval: (1) They transform images into captions or keywords, which are subsequently combined with the text query. (2) A text-retrieval system is then employed to acquire pertinent knowledge. This method, however, often leads to the omission of vital

Table 3.1 Evaluation of three proposed visual retrievers on OkVQA benchmark using **P**recision and **R**ecall: Caption-DPR achieves the highest Precision and Recall on all numbers of retrieved knowledge

Model	# of retrieved knowledge													
	1		5		10		20		50		80		100	
	P	R	P	R	P	R	P	R	P	R	P	R	P	R
BM25	37.63	37.63	35.21	56.72	34.03	67.02	32.62	75.90	29.99	84.56	28.46	88.21	27.69	89.91
Image-DPR	33.04	33.04	31.80	62.52	31.09	73.96	30.25	83.04	28.55	90.84	27.40	93.80	26.75	94.67
Caption-DPR	**41.62**	**41.62**	**39.42**	**71.52**	**37.94**	**81.51**	**36.10**	**88.57**	**32.94**	**94.13**	**31.05**	**96.20**	**30.01**	**96.95**

image details, such as contextual and background elements. Furthermore, caption generation models, when trained on specific domains, may not perform as well in varied real-world contexts. In response, the authors of [11] have developed a novel end-to-end Visual-Language Retriever (VL-Retriever), aptly named *ReViz*. This model is uniquely designed for "Reading and Visualizing" the query, utilizing the complete image rather than just focusing on object categories, keywords, or captions. ReViz integrates a Vision Transformer-based model, ViLT [72], for direct encoding of images from raw pixels, along with context inputs. It also incorporates BERT [29] to transform intricate, unstructured text into knowledge embeddings. The ReViz model distinguishes itself in two primary ways. Firstly, it negates the requirement for an additional cross-modal translator, such as a captioning model, or an object detector for representing images. Secondly, its end-to-end design facilitates the flexible retraining of its submodules, which can be crucial in bridging domain discrepancies. Encompassing both a multimodal query encoder and a knowledge encoder, the ReViz model excels in reading and visualizing input queries.

Next, we will describe ReViz architecture of their model. Additionally, the authors have developed a new benchmark, intended to augment the existing resources for multimodal-query retrieval. Alongside this, they have also designed pretraining tasks specific to multimodal-query retrieval, achieving commendable zero-shot performance.

3.4.5.1 Model Architecture

Multimodal Query Encoder. ReViz utilizes ViLT [72] to concurrently encode the text input T and image I. This marks a departure from most prior models that relied on CNNs for object feature detection. Instead, ViLT employs a patch projection technique to streamline image embeddings, using only 32×32 embeddings that require a mere 2.4 M parameters. This approach is significantly more efficient compared to earlier models that used ResNet or Fast-RCNN for feature embeddings. The process begins by dividing an image of size (HxW) into (H/PxW/P) patches, based on a predefined patch size ($P \times P$). Each patch is flattened and undergoes a linear transformation (through a fully connected layer) to become a D-dimensional vector, where D is the model's dimension. This yields a series of image vector sequences. Additionally, to compensate for the positional information loss during patch creation, positional embeddings are added to the patch embeddings. These embeddings are learnable parameters, enabling the model to comprehend the relative positions of the patches in the image. For text embeddings, the model straightforwardly utilizes the BERT tokenizer and embeddings. Ultimately, ViLT, functioning as a cross-encoder, takes the combined embeddings of images and text as input. The final multimodal representation is obtained by applying a linear projection and a hyperbolic tangent function to the first index token embedding.

$$\mathbf{Z}_q = \text{ViLT}(I, T) \tag{3.10}$$

3.4 Multimodal Retrieval

Knowledge Encoder. For encoding knowledge, the authors use a pre-trained BERT model [29]. This model generates a sequence of dense vectors $(h_1, \ldots, h_n)$ for each input token, with the final representation being the vector representation of the special token.

The authors employ a pre-trained BERT model [29] for knowledge encoding. This model produces a sequence of dense vectors corresponding to each input token $(h_1, \ldots, h_n)$. The final representation is derived from the vector representation of the special token [CLS].

$$\mathbf{Z}k = \text{BERT}(K) \tag{3.11}$$

Once the embeddings of the query and knowledge are calculated by the encoders, the relevancy score is derived by taking the inner-dot product of the embeddings.

$$\text{Score}(I, T, K) = \mathbf{Z}k^\top \cdot \mathbf{Z}_q \tag{3.12}$$

Training Procedure. The training objective of ReViz draws inspiration from the concept of instance discrimination, rooted in contrastive learning. The loss function, which is aimed at minimization, is outlined as follows:

$$\mathcal{L} = -\log \frac{\exp(\mathbf{z}q \cdot \mathbf{z}k)}{\exp(\mathbf{z}q \cdot \mathbf{z}k) + \sum_{\hat{\mathbf{k}} \in \mathbf{B}\mathbf{k}, \hat{\mathbf{k}} \neq \mathbf{k}} \exp(\mathbf{z}q \cdot \mathbf{z}\hat{\mathbf{k}})}, \tag{3.13}$$

In the equation above, $\mathbf{z}q$ represents the query embedding, $\mathbf{z}k$ stands for the relevant knowledge embedding, and $\mathbf{z}\hat{k}$ symbolizes the irrelevant knowledge embedding which acts as negative instances. All in-batch samples ($\mathbf{B_k}$) are utilized as the negative instances.

Training with Hard Negatives. The use of randomly selected samples as negative instances may lead to the creation of less-than-ideal metric spaces. To improve retrieval performance, the authors explore the use of hard negative sampling techniques.

To obtain significant hard negative samples, the initial step involves training ReViz under the guidance of Eq. 3.13. Subsequently, for each training question, they identify the top 100 knowledge instances (excluding the correct answer) to use as hard negative samples. It's crucial to note that this method of hard negative mining is employed solely in the fine-tuning phase for the downstream task, and not during the pretraining phase (which will be discussed in the subsequent section) (Fig. 3.2).

3.4.5.2 Multimodal Retriever Pretraining

Evidence suggests that pretraining a retriever on an unsupervised task closely related to retrieval can substantially boost the performance of downstream tasks [73–75]. Similarly, VL-ICT, a pretraining task for multimodal retrievers, takes inspiration from the ICT [74] task in the NLP field. Implementing VL-ICT has been shown to significantly enhance the performance of ReViz.

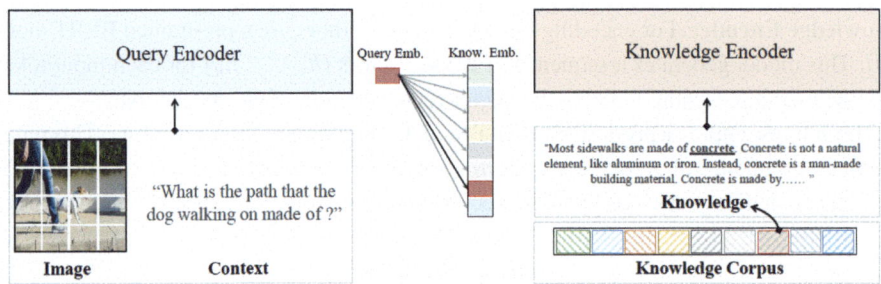

Fig. 3.2 Overall architecture of ReViz. ReViz consists of a Vision-Language Transformer that encodes the image and text and a knowledge encoder that projects the knowledge into knowledge embedding. During inference, ReViz selects the knowledge from the corpus that has the largest relevance score with the image-text embedding

The VL-ICT task integrates both visual and linguistic elements into the query. This task utilizes a training triplet, labeled as (I, T, K). In this triplet, I represents the visual input, and T is the linguistic input, each carrying unique information essential for the retrieval of knowledge. However, such a combination of I and T with distinct information is not typically found in natural settings. To address this, the authors propose an automated technique to create triplets that meet these specific criteria, which they will elaborate on in the following section.

VL-ICT Pretraining Data Collection. The methodology utilizes the Wikipedia-based Image Text (WiT) dataset as its core resource. Each entry in the WiT dataset includes a webpage title or an image caption, a passage, and an associated image. This image is used as the visual component I in the VL-ICT triplet. Web page titles or image captions often represent specific entities. These entities are matched with words in sentences found within the webpage passage that contain the title or caption. The sentences that match are designated as the text (T), while the rest of the passage, excluding the T sentences, serves as the knowledge (K). To guarantee that T and I provide unique yet essential information, any keywords in T that also appear in the caption are masked. Although the WiT dataset includes multilingual content, the VL-ICT process is applied exclusively to English entities within WiT. As a result of this procedure, a significant dataset comprising 3.2 million (I, T, K) training triplets has been created. This dataset serves as a robust foundation for both training and evaluation purposes (Fig. 3.3).

3.4.5.3 ReMuQ Dataset Creation

To expand the scope of research, the creator of ReViz [11] has developed ReMuQ, a new dataset tailored for multimodal-query retrieval. In ReMuQ, every query is matched with one definitive knowledge item. These queries are created by enhancing the WebQA ques-

3.4 Multimodal Retrieval

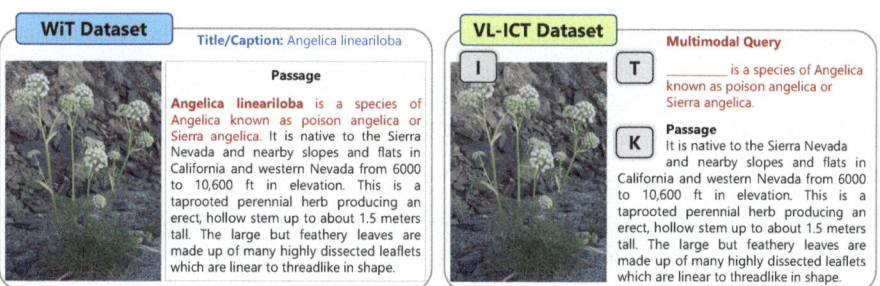

Fig. 3.3 Figure on the left shows an example of the WIT dataset [76], crawled from Wikipedia. The figure on the right shows our constructed (T, I, K) triplet: T is a sentence from the passage and the words overlapped with the title/caption is masked; K is the remaining passage after removing the sentence

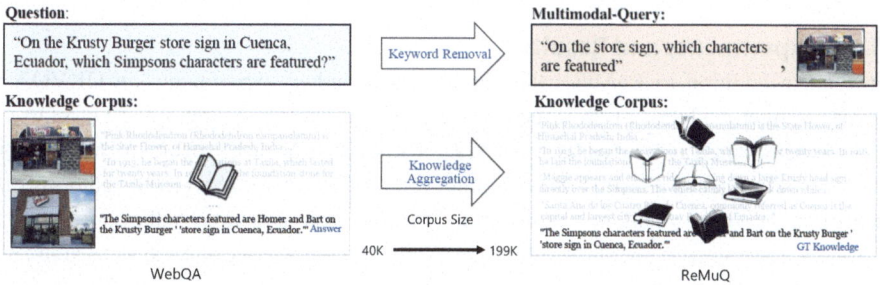

Fig. 3.4 Dataset creation procedure for ReMuQ using WebQA as the source of the raw data. The multimodal query in ReMuQ is the combination of an image and the question from WebQA where the overlapped information with the image is removed. The ground truth knowledge of ReMuQ is the answer from WebQA. The corpus consists of all answers and the distracted knowledge candidates given in ReMuQ

tions [77], accompanied by the assembly of a substantial corpus that serves as a knowledge repository for retrieval systems. WebQA is a diverse multimodalQA dataset encompassing a range of text question types, such as Yes/No, Open-ended (about aspects like shape or color), and multiple-choice (MC) questions. The dataset's images are obtained from Wikimedia Commons, with both the questions and their text responses generated by annotators. To formulate multimodal queries, the MC questions from WebQA are employed. These questions, associated with various choices in text or image form, act as knowledge sources. The correct answers could be exclusively text, exclusively image, or a fusion of both. Crucial steps have been implemented in the creation of these multimodal queries. The entire process of this creation, including key methodologies and stages, is detailed below and illustrated in Fig. 3.4.

(1) Question Filtering. Only multiple-choice questions that have answer choices containing both image and text are selected.

(2) Multimodal Query Construction. The initial multimodal query is the combination of the question and the corresponding image. To enforce a system to integrate information from both text and images, *tf-idf* is used to select keywords and then remove them from the question. The new multimodal query is then the concatenation of the augmented question and the image, with the text answer being the ground-truth knowledge.

(3) Retrieval Corpus Construction. The textual knowledge from all samples as the common knowledge corpus for multimodal retrieval is aggregated, resulting in a large corpus of ∼199 k knowledge descriptions.

(4) Dataset Train-Test Split. We divide ReMuQ into 70% for training and 30% as testing split. The new curated dataset contains 8418 training samples and 3609 testing samples, together with a knowledge corpus with 195, 837 knowledge descriptions.

3.4.5.4 Experiments and Results

Datasets. In addition to ReMuQ, the authors also conduct experiments on OKVQA to provide stronger evidence of their method's efficacy. Rather than employing OKVQA for its QA functionality, it is used as a testbed for the retrieval task, specifically to retrieve knowledge relevant to a question that contains the answer span. Moreover, the authors utilize two distinct corpora: a smaller corpus acquired from the Google search API as introduced in [11], and a larger corpus encompassing 21M Wikipedia knowledge entries, as used in [71].

Evaluation Metrics. Following the methodologies outlined in [11, 71], the authors assess the performance of their models using metrics such as Precision@K (P@K), Recall@K (R@K), and MRR@5. For the evaluation of the ReMuQ challenge, similar metrics are employed, with the exception that P@1 is used in place of P@5. This adjustment is made because ReMuQ is structured to have precisely one correct piece of knowledge for each query (Figs. 3.5 and 3.6).

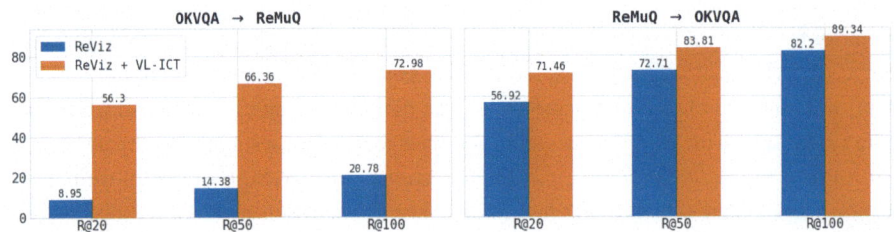

Fig. 3.5 Evaluation of out-of-domain performances of ReViz and ReViz + VL – ICT. For OKVQA, we retrieve knowledge from GS-112K corpus. VL-ICT substantially improves the generalization of ReViz. Other metrics are given in Appendix. X–>Y denotes using X as the training domain and Y as the testing domain

3.4 Multimodal Retrieval

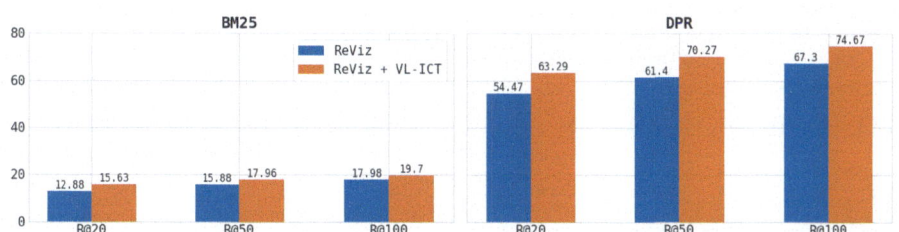

Fig. 3.6 Comparison of captioning-dependent retrievers using generated captions and ground truth captions. The ground truth captions always lead to better performance than generated caption

CLIP Baseline. CLIP [1] is a vision-language model pre-trained on over 400 M image-text pairs. We encode all knowledge descriptions via CLIP's textual encoder **K**. Then, given an image-text pair as the query, we use the image encoder to get the visual representations (**I**) and use the textual encoder to get the embedding of **Q**. We compute the inner-dot products between all encoded visual representations (**I**) and **K** to get the top-100 knowledge for evaluation, similarly for **Q**. Finally, we sum the scores and re-rank the top-100 knowledge.

CLIP [1] is a model that synergizes vision and language, pretrained on more than 400 million image-text pairs. In this application, all knowledge descriptions are encoded using CLIP's text encoder, denoted as **K**. For processing a query comprising an image-text pair, the image encoder is used to obtain visual representations (**I**), and the text encoder is employed for embedding the query text **Q**. The inner-dot product is computed between the encoded visual representations (**I**) and **K**, which is then used to identify the top-100 knowledge items for evaluation. A similar process is applied for **Q**. Finally, the scores from both processes are summed up, and the top-100 knowledge items are re-ranked based on these combined scores.

BM25 Baseline. BM25 [22] is a widely recognized and efficient algorithm for text-based retrieval tasks, relying on sparse representation. In this context, the image caption is utilized to represent the information contained in the image, thereby transforming the multimodal knowledge retrieval task into a purely text-based retrieval task.

DPR Baseline. The authors utilize DPR [3], which has been trained on the NaturalQuestions [83] dataset, as a baseline model for retrieving knowledge based on an input image-text pair. Initially, DPR's contextual encoder is used to index the corpus. Following this, the question and the image caption are merged to form a combined textual query. This query is then processed by the DPR's question encoder, which extracts its dense representation for subsequent calculations. The final step involves determining the most relevant knowledge items by calculating the inner-dot product between the query and the knowledge embeddings.

Results. Table 3.2 presents the performance metrics for both baseline models and ReViz, which has been pretrained on the VL-ICT task. Among the baseline models, DPR emerges as the strongest. Interestingly, despite CLIP's impressive performance in various classification

Table 3.2 Zero-shot performance of ReViz and baselines on two datasets: OKVQA and ReMuQ. OKVQA is evaluated on two knowledge sources. ReViz shows superior zero-shot performance in majority of the cases

Model	Dataset	KB-Size	Metric						
			MRR@5	P@5	R@5	R@10	R@20	R@50	R@100
CLIP-IMG+Q	OKVQA	GS-112 K	19.08	11.13	34.54	50.48	65.08	80.62	88.11
BM25 (GenCap)	OKVQA	GS-112 K	36.36	27.54	51.35	63.04	73.37	84.21	90.39
DPR (GenCap)	OKVQA	GS-112 K	39.15	27.72	55.56	66.44	75.59	87.17	92.42
ReViz+VL-ICT	OKVQA	GS-112 K	**45.77**	**33.18**	**64.05**	**75.39**	**84.21**	**91.64**	**94.59**
TRiG [71]	OKVQA	Wiki-21 M	–	–	45.83	57.88	72.11	80.49	86.56
CLIP-IMG+Q	OKVQA	Wiki-21 M	16.45	9.66	29.81	43.00	55.73	72.73	82.26
BM25 (GenCap)	OKVQA	Wiki-21 M	36.43	27.89	50.16	60.92	71.62	82.82	88.74
DPR (GenCap)	OKVQA	Wiki-21 M	41.15	28.10	59.41	71.13	81.73	89.90	93.39
ReViz+VL-ICT	OKVQA	Wiki-21 M	**44.03**	**32.94**	**62.43**	**73.44**	**82.28**	**89.93**	**93.76**
CLIP-IMG+Q	ReMuQ	199 K	0.34	0.17	0.78	1.36	2.41	7.34	47.88
BM25 (GenCap)	ReMuQ	199 K	3.80	5.59	8.78	10.75	12.88	15.88	17.98
DPR (GenCap)	ReMuQ	199 K	**31.23**	**35.79**	**43.42**	**48.77**	**54.47**	61.40	67.30
ReViz+VL-ICT	ReMuQ	199 K	23.61	29.52	39.43	46.77	53.56	**63.70**	**71.13**

and cross-modality pretraining tasks, it falls short in the multimodal query retrieval task. This indicates the complexity and challenges inherent in multimodal query retrieval for vision-language (VL) models. Crucially, ReViz demonstrates superior performance over all the baseline models in every metric for the OKVQA task, across both small and large corpus sizes. In the ReMuQ dataset, ReViz outperforms CLIP and BM25 in all metrics and surpasses DPR in two metrics. This highlights the effectiveness of the proposed pretraining task and the design of the ReViz model.

To further validate the efficacy of the VL-ICT pretraining task, the authors conduct an analysis of model performance after finetuning on downstream tasks. This comparison involves two variations of the ReViz model: (1) ReViz that is trained directly on the downstream task, and (2) ReViz initially pretrained on the VL-ICT task and subsequently finetuned on the downstream task. Additionally, the study encompasses two scenarios: in-domain and out-of-domain. In the in-domain scenario, a model is trained on the training set of a specific

3.4 Multimodal Retrieval

Table 3.3 Comparison of ReViz when it is fine-tuned on downstream tasks. We compare ReViz and ReViz+VL-ICT (our pretraining task). VL-ICT enables ReViz to be a stronger multimodal-query retrieval model

Model	Dataset	KB-Size	Metric						
			MRR@5	P@5	R@5	R@10	R@20	R@50	R@100
ReViz	OKVQA	GS-112K	46.92	34.51	66.05	77.80	86.33	93.34	95.90
ReViz+VL-ICT	OKVQA	GS-112K	**54.47**	**41.74**	**73.35**	**83.17**	**89.56**	**94.73**	**96.81**
ReViz	OKVQA	Wiki-21M	41.66	30.08	60.88	72.20	81.07	89.16	93.10
ReViz+VL-ICT	OKVQA	Wiki-21M	**43.68**	**31.36**	**61.91**	**72.63**	81.05	**89.28**	**93.44**
ReViz	ReMuQ	199K	41.03	49.08	62.40	71.63	78.92	86.60	92.17
ReViz+VL-ICT	ReMuQ	199K	**53.39**	**62.11**	**76.23**	**83.32**	**88.56**	**93.41**	**96.12**

domain X and then evaluated on the testing set of the same domain X. Conversely, in the out-of-domain scenario, the model is trained on the training set of domain X but evaluated on the testing set of a different domain Y. This approach helps in assessing the model's adaptability and performance across varying domains.

In-Domain Results. Table 3.3 displays the in-domain performance results. Across both datasets, ReViz that has undergone pretraining consistently surpasses the performance of the vanilla (non-pretrained) ReViz. This trend indicates that the pretraining task enhances ReViz's ability to align multimodal queries more effectively with the corresponding relevant knowledge.

Out-of-Domain Results. In this investigation, the author explores the effectiveness of the VL-ICT pretraining task in improving ReViz's generalization capability. Two scenarios are considered: training on OKVQA (domain **X**) and testing on ReMuQ (domain **Y**), as well as the reverse configuration. As presented in Table 3.5, it is evident that ReViz + VL-ICT + **X** consistently outperforms ReViz+**X** on **Y**, particularly when **X** corresponds to OKVQA and **Y** to ReMuQ. This observation suggests that models pretrained with VL-ICT tasks exhibit enhanced robustness compared to those without VL-ICT. Furthermore, it's worth noting that despite the improvements in generalization achieved through VL-ICT pretraining, there remains a substantial performance gap compared to fine-tuning. This underscores the distinct nature of the OKVQA and ReMuQ tasks, emphasizing the importance of ReMuQ as a valuable complement to the study of multimodal query retrieval tasks. In the comparative analysis of ReViz with existing retrieval methods for the OKVQA task, it's important to note that many models on the OKVQA leaderboard primarily report question answering accuracy rather than retrieval performance. However, the experiments include systems that provide insights into the retrieval aspect of OKVQA performance (Table 3.4).

Table 3.4 Comparison of our best model with existing models on OKVQA. "FT" denotes fine-tuning. Our model surpasses existing methods by significant margins with or without fine-tuning and with different knowledge corpus

Model	FT	KB-Size	Metric						
			MRR@5	P@5	R@5	R@10	R@20	R@50	R@100
VRR-IMG [11]	✓	GS-112 K	–	31.80	62.52	73.96	83.04	90.84	94.67
VRR-CAP [11]	✓	GS-112 K	–	39.42	71.52	81.51	88.57	94.13	96.95
ReViz+VL-ICT	✓	GS-112 K	54.47	41.74	73.35	83.17	89.56	94.73	96.81
TRiG [71]	×	Wiki-21 M	–	–	45.83	57.88	72.11	80.49	86.56
ReViz+VL-ICT	×	Wiki-21 M	44.03	32.94	62.43	73.44	82.28	89.93	93.76

Fine-tune Model Baselines. Luo et al. [11] present two fine-tuned multimodal retrievers: VRR-IMG which uses LXMERT [81] and VRR-CAP to convert the image into captions for knowledge retrieval. Both retrievers use GS-112K as the knowledge corpus. TriG [71] uses zeroshot retriever and Wikipedia 21 M as the knowledge corpus. Since these systems use either fine-tuned retriever or zero-shot retrievers, for fair comparison, we compare the best fine-tuned model and zeroshot model with the corresponding corpus.

Fine-tune Results. In the fine-tuning scenario, with the exception of R@100, the author's models consistently outperform previous methods across all metrics. Similarly, in the zero-shot case, the author's model surpasses previous models on all metrics by significant margins.

In VL-ICT, the author's approach involves masking keywords in sentences to prevent information leakage. However, it has been observed that certain masked sentences still overlap with the retrieved knowledge. This overlap may make the VL-ICT task easier, potentially diminishing the benefits of pre-training. To investigate the optimal mask ratio, experiments were conducted by randomly masking words in sentences at different ratios. These experiments were conducted on a smaller corpus of 1 million VL-ICT training triplets, and models were trained for one epoch. Figure 3.7 illustrates the results, showing that removing 20% of the keywords yielded the best performance among all ratios and outperformed keeping the sentences intact (0% masking).

Previous systems relying on caption generation models can be impacted by the quality of generated captions, which may hinder retrieval performance, especially when the caption generation model is not trained on the same domain as the downstream task. In the ReMuQ dataset, the images are sourced from Wikipedia, while the caption generator is trained on MS-COCO [5]. To assess this, the author compared two baselines, BM25 and DPR, using both ground-truth image captions and generated captions. Table 3.6 clearly demonstrates that using ground-truth captions significantly outperforms the use of generated captions in all cases. This indicates that the caption generator plays a crucial role in converting image information into captions and highlights the limitations of previous methods, providing justification for the exploration of end-to-end training.

3.5 Downstream Tasks Using MMIR

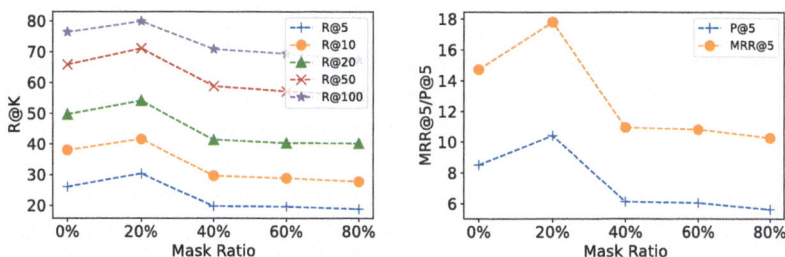

Fig. 3.7 Effect of the masking ratio of sentences in VL-ICT task on ReViz's performance on OKVQA Task. We use GS112K as the knowledge corpus

3.4.5.5 Conclusion

In their work, Luo et al. [11] introduce knowledge retrieval using multimodal queries encompassing both vision and language. This domain, in comparison to established retrieval tasks, presents a greater challenge and remains relatively unexplored. Moreover, the concept of multimodal-query information retrieval holds immense potential for a wide range of applications, extending beyond the conventional image, text, and video retrieval to areas like question-answering, recommendation systems, and personal assistants. To facilitate advancements in these domains, the authors introduce the ReMuQ dataset, tailored to support the development of such functionalities. In their approach, they propose an end-to-end VL-retriever model named ReViz, notable for its independence from intermediary image-to-text translation modules. Additionally, they introduce a novel weakly-supervised task called VL-ICT, designed for large-scale pre-training. Through extensive evaluations of ReMuQ and OK-VQA datasets, Luo et al. [11] demonstrates that ReViz stands out among all retrieval models in both zero-shot and fine-tuning scenarios. Their dataset and model not only contribute to the current state of research but also serve as a solid foundation for future endeavors, potentially leading to groundbreaking discoveries and innovative applications in the realm of multimodal-query information retrieval.

3.5 Downstream Tasks Using MMIR

Over the last decade, there have been remarkable advancements in both image and text understanding. These advancements encompass a wide range of tasks: for images, these include scene understanding, image classification, object and action detection, semantic segmentation, and 3D reconstruction; for text, they involve entailment detection, sentiment analysis, question answering, and semantic similarity assessments. The success in these areas can be largely attributed to the availability of extensive training datasets and advanced machine-learning algorithms capable of modeling high-dimensional functions. More recently, there has been a growing interest in exploring the interplay between vision and language. This isn't just to develop more robust and well-grounded models, but also to forge a new paradigm in 'multimodal learning'. This involves tasks that necessitate an understanding of both visual and textual inputs. The field of vision and language (V&L), which focuses on tasks involving both visual and textual elements, has seen significant growth. The complex interplay of perception and semantics in these tasks has spurred renewed interest in linking observable entities in scenes with facts, world knowledge, and human concepts.

This section will review works in the V&L domain that incorporate external knowledge, highlighting how a multimodal retriever can be instrumental. We will discuss the significance of external knowledge and delve into three major downstream tasks that require Multi-Modal Information Retrieval (MMIR).

The Role of Knowledge. What is knowledge and how can it be acquired? This simple question has been debated and discussed for millennia by philosophers. The school of philosophy that is perhaps the closest to modern discourse about knowledge in computer science and machine learning, is Advaita Vedanta, which accepts six means of acquiring knowledge: perception, inference, analogy, implication, negative proof, and testimony by reliable experts [78]. For V&L tasks, perception has been well-studied via feature extraction using object detectors [79] for representing images and using vectors as representations for words and sentences [24, 26]. For inference, several neural-network models have been designed for V&L tasks [80, 81]. Through perception and inference, models can perform tasks that require understanding the observable visual and textual entities in a scene. However, investigating the role of knowledge sources and methods beyond perception and inference in V&L is still a nascent research topic. In this survey, we will denote this by *external knowledge (XK)*, i.e., knowledge that is not perceivable from the inputs.

Consider the example in Fig. 3.8. A visual question answering model when asked "*How many people are in this image?*" has enough information in the image to predict the answer "*3*". However, answering questions such as "*How many chromosomes do these creatures have?*" requires not only detecting humans in this image but also accessing the scientific fact that humans have 23 chromosomes. This piece of knowledge is not perceivable and might only be found in scientific articles, textbooks, or encyclopedias. Similarly, perception-based image captioning models may predict the caption "*a person giving a speech*" while knowledge-based captioners may connect this perception with the names of people and known locations and predict "*Barack Obama giving a speech in the US Congress in front of the American flag*". Although large datasets can be collected with millions of images and textual annotations, it is unlikely that the dataset will contain the necessary knowledge to improve V&L performance for every context and every image. More importantly, the existing paradigm of learning the mapping between inputs and expected outputs in an end-to-end fashion limits the distillation of such contextual, factual, and scientific knowledge and drastically affects performance on novel, rare, and long-tail concepts. These examples motivate the need for using external knowledge in V&L tasks. Thus we need methods that can retrieve and leverage relevant external knowledge, given the visual and textual inputs.

3.5 Downstream Tasks Using MMIR

Perception-based VQA:
How many people are in this image? → 3
Knowledge-based VQA:
How many chromosomes do these creatures have? → 23

Perception-based Caption:
A person giving a speech.
Knowledge-based Caption:
Barack Obama giving a speech in the US Congress in front of the American flag.

Fig. 3.8 Examples of V&L tasks that motivate the need for external knowledge. The role of external knowledge is to connect perceptible properties such as objects, colors, and shapes, with various kinds of knowledge that help us understand the image better—these may include world knowledge (such as names of people and places), scientific facts, or commonsense knowledge (based on behavior and internal states of people)

We categorize the role of external knowledge in vision-and-language tasks into three major paradigms. The **first paradigm** involves the creation of new tasks that cannot be learned using standard existing datasets, but require collection of data from knowledge sources. For instance, a new task of captioning images and videos has emerged where the output is expected to describe the intentions of people in performing actions and events before and after the action. Existing datasets such as MS-COCO [5] do not contain such annotations, and therefore, knowledge bases about human internal states and behaviors need to be leveraged. The **second paradigm** involves algorithmic methods, such as retrieval, learning, or reasoning methods). These methods are used either when the information contained in the input is insufficient to solve the task (VQA example in Fig. 3.8 or to improve the quality of the output (captioning example in Fig. 3.8). The **third paradigm** deals with evaluation metrics designed using external knowledge about a task. For instance, from the rules of linguistics, it is known that when negation is added to questions with yes/no answers, the answer should also be negated [82]. Thus evaluation protocols, guided by such domain knowledge can be designed to test model robustness and consistency. In this survey, we will discuss work on image captioning, visual question answering, visual dialog, and vision-language navigation, for these three paradigms.

Such data surround humans every day due to the digital world that we live in, and the rise of smartphones, social media platforms, and the Internet of Things (IoT) have led to a wealth of user-generated content, consisting of diverse data types. Therefore, understanding and utilizing multimodal data is an urgent need for both academic and industry research. There are numerous amount of research topics on multimodal data, such as visual question answering, image captioning, and image summarization, and more recent developments in image generation based on text description. These can be classified as classification tasks or generation tasks. Information retrieval can also be considered as a classification task with unique differences in that the classification space is much larger than other classification tasks like the VQA, and also the classes could be dynamically growing as new information comes in. In the following, we summarize two major tasks where external knowledge is required and thus an MMIR system is needed.

3.5.1 VQA

Visual Question Answering requires a joint understanding of visual and language information to predict an answer to the question. Different from the conventional VQA datasets, such as VQA [84], where the answer can be reached by the content in the image (e.g., the question is how many people are in the image), in knowledge involved VQA, the given inputs are not enough to reach the answer and require external knowledge to complete the task. In general, there are two types of questions, multiple choices where a system is given a set of candidates and supposed to select one correct answer. On the other hand, open-ended questions do not provide answer choices and the answer is free-form text.

Image Question Answering Visual7W-telling dataset [85] contains 328 K multi-choice visual questions of various types (What, Where, When, Who, Why, and How), where lots of diverse questions involve external commonsense knowledge. FVQA [86] focuses on commonsense knowledge. KB-VQA [87] are generated from templates using the entities in the image and the concept about the entities in DBpedia. KVQA [88] requires world knowledge about named entities (e.g. Barack Obama) in the image. Text-KVQA [89] is about the text in the image and the knowledge of such text. OKVQA [8] is less entity-centric and includes 10 types of external knowledge, e.g., knowledge about weather and cooking. KRVQA [90] focuses on multi-hop reasoning using the knowledge from VisualGenome and the knowledge base. WEBQA [10] aims for aggregating multi-source information (multi-hop) and the form of the external knowledge can be both image and text. In addition, previous datasets are associated with short answers, while WEBQA has long answers. The source of the images includes Microsoft COCO [5], ImageNet [19], and Wikimedia.

Video Question Answering Compared to images, video contains more temporal and spatial information, and thus more complex questions (e.g., *How question*) can be asked. In addition, video QA requires reasoning over the dialogues and actions shown in specific scenes and relating them to the overall storyline already seen. For example, to answer questions in TVQA+ [91], systems need to retrieve relevant clips in a long video and understand the dialogue but do not require external knowledge, and many video question answer datasets do not require external knowledge. On the other hand, VCR [92] requires commonsense knowledge and asks for the rationale to explain why the answer is right. This task requires higher-order cognition and commonsense reasoning about the world. KnowIT VQA [93] requires knowledge of the TV show The Big Bang Theory. KnowIT-X VQA [94] is similar to KnowIT but developed for another TV show "Friends". Due to the different domains of the two TV shows, KnowIT focuses more on scientific knowledge while KnowIT-X is more about the relationships among characters. The source of the video includes TV shows, Large Scale Movie Description Challenge, and YouTube videos.

3.5.2 Caption Generation

Captioning generation, as an important VL task, requires deep interpretation of the visual observation. While promising advances are presented in existing efforts, most of them translate the visual input into human-readable descriptions directly without exploiting external knowledge. Notably, certain captioning challenge under specific domains requires the incorporation of external knowledge due to the limitation of the ground-truth caption. This is mainly reflected in two aspects: (1). Producing captions with novel object/concept. Hendricks et al. [95] proposes to collect object knowledge from external corpora, e.g., Wikipedia, and British National Corpus, to facilitate novel object captioning. Similar works also include [96–98], where the specific named entities are generated by retrieving from the relevant texts. Whitehead et al. [99] propose to retrieve topically related news documents and extract the events and named entities from these documents for video captioning tasks. Some work [100, 101] employs the copying mechanism where the novel objects are selected from the external corpora. Likewise, Krishnamoorthy et al. [102] incorporate text-mined SVO for video captioning. A very recent work [103] for the first time, proposes to use a multimodal retriever (e.g., CLIP [1]) to obtain long-tailed concepts from keyword vocabulary as external knowledge. (2). Inferring the contextual situations or rationale. There also exist works leveraging knowledge graphs to represent the object entity relations to speculate the hidden observations to assist captions generation [104, 105], or injecting prior knowledge [106] that uses a set of latent topics as anchors to produce highly probable words with low data regime. Notably, some captioning datasets involve inferring annotations that require the model to relate or generate knowledge from the visual observation. Fang et al. [107] aims to generate the latent commonsense knowledge given the observable actions in videos. Sidorov et al. [108] propose the challenge to recognize the text and relate it to its visual context, requiring visual reasoning between multiple text tokens and visual entities.

3.6 Evaluation

In this section, we introduce the widely used evaluation metrics for multimodal retrieval. There are some common terms across different metrics:

- True Positives (TP) represent the number of positive instances that are correctly identified as positive.
- False Positives (FP) represent the number of negative instances that are incorrectly identified as positive.

- False Negatives (FN) represent the number of positive instances that are incorrectly identified as negative.

We will assume a scenario where the corpus has 100K items, and a query Q has 50 relevant items in this corpus. System A retrieves 100 documents, and 30 of them are relevant.

Precision is the proportion of retrieved documents that are relevant. Mathematically, it is expressed as below,

$$\text{Precision} = \frac{TP}{TP + FP}. \quad (3.14)$$

The Precision value for System A is 30/100 = 0.3.

Recall is the proportion of relevant documents that are retrieved. Mathematically, it is expressed as below,

$$\text{Recall} = \frac{TP}{TP + FN} \quad (3.15)$$

The recall value of System A is 30/50 = 0.6.

However, the above calculation does not take the number of items that systems can retrieve. Image that system B retrieves 200 items and 30 of them are relevant. By Eq. 3.15, then System A and B both have a recall of 0.6, however, system A might be more efficient than B because it retrieves fewer items and has the same amount of relevant items. To take the number of retrieved items into account, Recall@K (or Precision@K) is used in most of the literature such that each system can retrieve K items.

F1 Score is the harmonic mean of precision and recall and provides a single metric that balances both values. It's particularly useful when two systems that may have different precision and recall trade-offs. Mathematically, it is expressed as below,

$$F1 = 2 \cdot \frac{\text{Precision} \cdot \text{Recall}}{\text{Precision} + \text{Recall}} \quad (3.16)$$

Mean Average Precision (MAP) is measuring the average precision scores after each relevant document is retrieved. It's particularly used in tasks where the order of the documents is important, such as in ranked retrieval results.

$$AP = \frac{1}{m} \sum_{k=1}^{n} (\text{Precision@k} \cdot \text{Recall@k}) \quad (3.17)$$

$$MAP = \frac{1}{Q} \sum_{q=1}^{Q} (AP_q) \quad (3.18)$$

where m is the number of relevant document, n is the number of retrieved documents, Q is the number of questions.

3.6 Evaluation

Normalized Discounted Cumulative Gain (NDCG) is another metric used in ranked retrieval results that takes into account the position of the retrieved documents in the result list. It gives higher importance to relevant documents appearing at the top of the list. A mathematical expression is as follows,

$$DCG_p = \sum_{i=1}^{p} \frac{rel_i}{\log_2(i+1)} \quad (3.19)$$

$$IDCG_p = \sum_{i=1}^{p} \frac{rel_i^{ideal}}{\log_2(i+1)} \quad (3.20)$$

$$NDCG_p = \frac{DCG_p}{IDCG_p} \quad (3.21)$$

Where rel_i is the relevance of the result at position i,

Reciprocal Rank is the multiplicative inverse of the rank of the first relevant document. For example, if the first relevant document is at position 3, the reciprocal rank is 1/3.

Ranking Percentile refers to the percentage of a specific subset of data that a particular data point outperforms. This can be used to measure how well a particular item or result ranks compared to all other items or results. For instance, if a document ranks at the 90th percentile, it performs better than 90% of the other documents in the set. This is particularly useful for understanding the relative position of an item in a list, which is critical for tasks such as search engine ranking, recommendation systems, and other ranking problems.

$$\text{Percentile rank} = \frac{\text{Number of values less than 'x'}}{\text{Total number of values in the dataset}} \times 100 \quad (3.22)$$

Hit@K A "hit" refers to a document or item returned by a system that is relevant to the user's query. And Hit@K refers to the ratio of relevant items in the top K retrieved items. Here's a simplified example: Suppose K is equal to 10, and among the 10 documents retrieved by an IR system, 7 of them are relevant to the query. In this case, the "hit rate" would be 70% (7 out of 10). Mathematically,

$$\text{Hit@K} = \frac{1}{K} \sum_{i=1}^{K} I(\text{rel}(d_i) = 1). \quad (3.23)$$

This metric has been used in many multimodal retriever benchmarks.

Advanced Evaluation For open-ended VQA tasks, above mentioned standard accuracy metrics can be too stringent as they require a predicted answer to exactly match the ground-truth answer. This issue is addressed in literature in two ways: one, by collecting multiple human-annotated answers and second, automatic evaluation using soft matching metrics such as edit distances, Wu-Palmer similarity (WUPS) [109], Alter- native Answer Sets (AAS) [110].

Challenges in Evaluation One big problem of the evaluation is the false positive, i.e. potentially relevant results are considered false because they are not presented in the ground truth. This is due to the limited annotations. The retrieval corpus is usually too large to label to ground truth of each document w.r.t to a query. To optimize human effort, a set of candidates is initially retrieved for each query, and then annotators are requested to label each candidate in the set. However, this approach may result in potential correct answers being excluded from the candidate set, causing them to be deemed false or unrelated to the query.

3.7 Broader Impact of Multimodal Retrieval

The advent and refinement of multimodal retrieval technologies, which integrate and leverage diverse data types like text, images, and audio, are having profound impacts across various sectors. These technologies, much like their generative counterparts, have evolved from being niche research topics to becoming integral components of mainstream applications. Their ability to simultaneously analyze and retrieve information from different modalities has revolutionized how we interact with and process digital content. In sectors like e-commerce, healthcare, and digital libraries, multimodal retrieval systems have enhanced user experience and operational efficiency. For instance, in e-commerce, these systems allow users to search for products using both text and images, leading to more accurate and satisfying shopping experiences. In healthcare, the integration of textual patient records with medical imagery facilitates more comprehensive and faster diagnoses.

However, the deployment of multimodal retrieval systems is not without challenges. One major concern is the balance between accuracy and computational efficiency, especially when dealing with large and diverse datasets. There's also the issue of ensuring fairness and avoiding biases in retrieval results, as these systems often rely on historical data that may contain inherent biases. Privacy concerns are another critical aspect, particularly when handling sensitive data like personal images or health records. Ensuring the security and privacy of such data is paramount to maintain user trust and comply with regulatory standards.

As we look to the future, the potential for multimodal retrieval systems to replace or augment certain tasks is significant. For instance, in scenarios where approximate results are acceptable, these systems could efficiently provide relevant content without the need for exact matches, as seen in the use of generative models for image generation. The key to realizing the full potential of these technologies lies in addressing their current limitations through continuous research, the development of robust and fair algorithms, and the implementation of stringent privacy safeguards. In conclusion, multimodal retrieval technologies, while still evolving, offer vast possibilities for improving how we search for and interact with information across various domains. Their broader impact is contingent on our ability to refine these systems for greater accuracy, fairness, and efficiency, ensuring they serve as valuable tools in our increasingly digital world.

References

1. Alec Radford, Jong Wook Kim, Chris Hallacy, Aditya Ramesh, Gabriel Goh, Sandhini Agarwal, Girish Sastry, Amanda Askell, Pamela Mishkin, Jack Clark, Gretchen Krueger, and Ilya Sutskever. Learning transferable visual models from natural language supervision. In Marina Meila and Tong Zhang, editors, *Proceedings of the 38th International Conference on Machine Learning, ICML 2021, 18–24 July 2021, Virtual Event*, volume 139 of *Proceedings of Machine Learning Research*, pages 8748–8763. PMLR, 2021. URL http://proceedings.mlr.press/v139/radford21a.html.
2. Liunian Harold Li, Pengchuan Zhang, Haotian Zhang, Jianwei Yang, Chunyuan Li, Yiwu Zhong, Lijuan Wang, Lu Yuan, Lei Zhang, Jenq-Neng Hwang, et al. Grounded language-image pretraining. In *Proceedings of the IEEE/CVF Conference on Computer Vision and Pattern Recognition*, pages 10965–10975, 2022a.
3. Vladimir Karpukhin, Barlas Oguz, Sewon Min, Patrick Lewis, Ledell Wu, Sergey Edunov, Danqi Chen, and Wen-tau Yih. Dense passage retrieval for open-domain question answering. In *Proceedings of the 2020 Conference on Empirical Methods in Natural Language Processing (EMNLP)*, pages 6769–6781, Online, 2020a. Association for Computational Linguistics. https://doi.org/10.18653/v1/2020.emnlp-main.550. URL https://aclanthology.org/2020.emnlp-main.550.
4. Ting Chen, Simon Kornblith, Mohammad Norouzi, and Geoffrey E. Hinton. A simple framework for contrastive learning of visual representations. In *Proceedings of the 37th International Conference on Machine Learning, ICML 2020, 13–18 July 2020, Virtual Event*, volume 119 of *Proceedings of Machine Learning Research*, pages 1597–1607. PMLR, 2020. URL http://proceedings.mlr.press/v119/chen20j.html.
5. Tsung-Yi Lin, Michael Maire, Serge Belongie, James Hays, Pietro Perona, Deva Ramanan, Piotr Dollár, and C Lawrence Zitnick. Microsoft coco: Common objects in context. In *European conference on computer vision*, pages 740–755. Springer, 2014.
6. Peter Young, Alice Lai, Micah Hodosh, and Julia Hockenmaier. From image descriptions to visual denotations: New similarity metrics for semantic inference over event descriptions. *Transactions of the Association for Computational Linguistics*, 2:67–78, 2014a. https://doi.org/10.1162/tacl_a_00166. URL https://aclanthology.org/Q14-1006.
7. Javier Marın, Aritro Biswas, Ferda Ofli, Nicholas Hynes, Amaia Salvador, Yusuf Aytar, Ingmar Weber, and Antonio Torralba. Recipe1m+: A dataset for learning cross-modal embeddings for cooking recipes and food images. *IEEE Transactions on Pattern Analysis and Machine Intelligence*, 43(1):187–203, 2021.
8. Kenneth Marino, Mohammad Rastegari, Ali Farhadi, and Roozbeh Mottaghi. OK-VQA: A visual question answering benchmark requiring external knowledge. In *IEEE Conference on Computer Vision and Pattern Recognition, CVPR 2019, Long Beach, CA, USA, June 16–20, 2019*, pages 3195–3204. Computer Vision Foundation / IEEE, 2019a https://doi.org/10.1109/CVPR.2019.00331. URL http://openaccess.thecvf.com/content_CVPR_2019/html/Marino_OK-VQA_A_Visual_Question_Answering_Benchmark_Requiring_External_Knowledge_CVPR_2019_paper.html.

9. Dustin Schwenk, Apoorv Khandelwal, Christopher Clark, Kenneth Marino, and Roozbeh Mottaghi. A-okvqa: A benchmark for visual question answering using world knowledge. *ArXiv preprint*, arXiv:abs/2206.01718, 2022. URL https://arxiv.org/abs/2206.01718.
10. Yingshan Chang, Guihong Cao, Mridu Narang, Jianfeng Gao, Hisami Suzuki, and Yonatan Bisk. Webqa: Multihop and multimodal QA. In *IEEE/CVF Conference on Computer Vision and Pattern Recognition, CVPR 2022, New Orleans, LA, USA, June 18–24, 2022*, pages 16474–16483. IEEE, 2022b. https://doi.org/10.1109/CVPR52688.2022.01600. URL https://doi.org/10.1109/CVPR52688.2022.01600.
11. Man Luo, Zhiyuan Fang, Tejas Gokhale, Yezhou Yang, and Chitta Baral. End-to-end knowledge retrieval with multi-modal queries. *ArXiv preprint*, arXiv:abs/2306.00424, 2023a. URL https://arxiv.org/abs/2306.00424.
12. Nam Vo, Lu Jiang, Chen Sun, Kevin Murphy, Li-Jia Li, Li Fei-Fei, and James Hays. Composing text and image for image retrieval - an empirical odyssey. In *IEEE Conference on Computer Vision and Pattern Recognition, CVPR 2019, Long Beach, CA, USA, June 16–20, 2019*, pages 6439–6448. Computer Vision Foundation / IEEE, 2019. https://doi.org/10.1109/CVPR.2019.00660. URL http://openaccess.thecvf.com/content_CVPR_2019/html/Vo_Composing_Text_and_Image_for_Image_Retrieval_-_an_Empirical_CVPR_2019_paper.html.
13. Hui Wu, Yupeng Gao, Xiaoxiao Guo, Ziad Al-Halah, Steven Rennie, Kristen Grauman, and Rogério Feris. Fashion IQ: A new dataset towards retrieving images by natural language feedback. In *IEEE Conference on Computer Vision and Pattern Recognition, CVPR 2021, virtual, June 19–25, 2021*, pages 11307–11317. Computer Vision Foundation / IEEE, 2021a. https://doi.org/10.1109/CVPR46437.2021.01115. URL https://openaccess.thecvf.com/content/CVPR2021/html/Wu_Fashion_IQ_A_New_Dataset_Towards_Retrieving_Images_by_Natural_CVPR_2021_paper.html.
14. JaeYun Lee and Incheol Kim. Vision–language–knowledge co-embedding for visual commonsense reasoning. *Sensors (Basel, Switzerland)*, 21(9), 2021.
15. Violetta Shevchenko, Damien Teney, Anthony Dick, and Anton van den Hengel. Reasoning over vision and language: Exploring the benefits of supplemental knowledge. In *Proceedings of the Third Workshop on Beyond Vision and LANguage: inTEgrating Real-world kNowledge (LANTERN)*, pages 1–18, Kyiv, Ukraine, 2021. Association for Computational Linguistics. URL https://aclanthology.org/2021.lantern-1.1.
16. Ranjay Krishna, Yuke Zhu, Oliver Groth, Justin Johnson, Kenji Hata, Joshua Kravitz, Stephanie Chen, Yannis Kalantidis, Li-Jia Li, David A Shamma, et al. Visual genome: Connecting language and vision using crowdsourced dense image annotations. *International Journal of Computer Vision*, 123(1):32–73, 2017.
17. Christiane Fellbaum. Wordnet. In *Theory and applications of ontology: computer applications*, pages 231–243. Springer, 2010.
18. Houda Alberts, Ningyuan Huang, Yash Deshpande, Yibo Liu, Kyunghyun Cho, Clara Vania, and Iacer Calixto. VisualSem: a high-quality knowledge graph for vision and language. In *Proceedings of the 1st Workshop on Multilingual Representation Learning*, pages 138–152, Punta Cana, Dominican Republic, 2021. Association for Computational Linguistics. https://doi.org/10.18653/v1/2021.mrl-1.13. URL https://aclanthology.org/2021.mrl-1.13.
19. Jia Deng, Wei Dong, Richard Socher, Li-Jia Li, Kai Li, and Fei-Fei Li. Imagenet: A large-scale hierarchical image database. In *2009 IEEE Computer Society Conference on Computer Vision and Pattern Recognition (CVPR 2009), 20-25 June 2009, Miami, Florida, USA*, pages 248–255. IEEE Computer Society, 2009. https://doi.org/10.1109/CVPR.2009.5206848.
20. Man Luo. Neural retriever and go beyond: A thesis proposal. In *Proceedings of the 2022 Conference of the North American Chapter of the Association for Computational Linguistics: Human Language Technologies: Student Research Workshop*, pages 59–67, 2022.

21. Gerard Salton and Christopher Buckley. Term-weighting approaches in automatic text retrieval. *Information processing & management*, 24(5):513–523, 1988.
22. Stephen Robertson, Hugo Zaragoza, et al. The probabilistic relevance framework: Bm25 and beyond. *Foundations and Trends® in Information Retrieval*, 3(4):333–389, 2009.
23. Ashish Vaswani, Noam Shazeer, Niki Parmar, Jakob Uszkoreit, Llion Jones, Aidan N. Gomez, Lukasz Kaiser, and Illia Polosukhin. Attention is all you need. In Isabelle Guyon, Ulrike von Luxburg, Samy Bengio, Hanna M. Wallach, Rob Fergus, S. V. N. Vishwanathan, and Roman Garnett, editors, *Advances in Neural Information Processing Systems 30: Annual Conference on Neural Information Processing Systems 2017, December 4–9, 2017, Long Beach, CA, USA*, pages 5998–6008, 2017a. URL https://proceedings.neurips.cc/paper/2017/hash/3f5ee243547dee91fbd053c1c4a845aa-Abstract.html.
24. Jacob Devlin, Ming-Wei Chang, Kenton Lee, and Kristina Toutanova. BERT: Pre-training of deep bidirectional transformers for language understanding. In *Proceedings of the 2019 Conference of the North American Chapter of the Association for Computational Linguistics: Human Language Technologies, Volume 1 (Long and Short Papers)*, pages 4171–4186, Minneapolis, Minnesota, 2019b. Association for Computational Linguistics. https://doi.org/10.18653/v1/N19-1423. URL https://aclanthology.org/N19-1423.
25. Tomas Mikolov, Kai Chen, Greg Corrado, and Jeffrey Dean. Efficient estimation of word representations in vector space. *arXiv preprint* arXiv:1301.3781, 2013.
26. Jeffrey Pennington, Richard Socher, and Christopher Manning. GloVe: Global vectors for word representation. In *Proceedings of the 2014 Conference on Empirical Methods in Natural Language Processing (EMNLP)*, pages 1532–1543, Doha, Qatar, 2014. Association for Computational Linguistics. https://doi.org/10.3115/v1/D14-1162. URL https://aclanthology.org/D14-1162.
27. Sepp Hochreiter and Jürgen Schmidhuber. Long short-term memory. *Neural computation*, 9(8):1735–1780, 1997.
28. Junyoung Chung, Caglar Gulcehre, KyungHyun Cho, and Yoshua Bengio. Empirical evaluation of gated recurrent neural networks on sequence modeling. *arXiv preprint* arXiv:1412.3555, 2014.
29. Jacob Devlin, Ming-Wei Chang, Kenton Lee, and Kristina Toutanova. BERT: Pre-training of deep bidirectional transformers for language understanding. In *Proceedings of the 2019 Conference of the North American Chapter of the Association for Computational Linguistics: Human Language Technologies, Volume 1 (Long and Short Papers)*, pages 4171–4186, Minneapolis, Minnesota, 2019a. Association for Computational Linguistics. https://doi.org/10.18653/v1/N19-1423. URL https://aclanthology.org/N19-1423.
30. Tom B. Brown, Benjamin Mann, Nick Ryder, Melanie Subbiah, Jared Kaplan, Prafulla Dhariwal, Arvind Neelakantan, Pranav Shyam, Girish Sastry, Amanda Askell, Sandhini Agarwal, Ariel Herbert-Voss, Gretchen Krueger, Tom Henighan, Rewon Child, Aditya Ramesh, Daniel M. Ziegler, Jeffrey Wu, Clemens Winter, Christopher Hesse, Mark Chen, Eric Sigler, Mateusz Litwin, Scott Gray, Benjamin Chess, Jack Clark, Christopher Berner, Sam McCandlish, Alec Radford, Ilya Sutskever, and Dario Amodei. Language models are few-shot learners. In Hugo Larochelle, Marc'Aurelio Ranzato, Raia Hadsell, Maria-Florina Balcan, and Hsuan-Tien Lin, editors, *Advances in Neural Information Processing Systems 33: Annual Conference on Neural Information Processing Systems 2020, NeurIPS 2020, December 6–12, 2020, virtual*, 2020b. URL https://proceedings.neurips.cc/paper/2020/hash/1457c0d6bfcb4967418bfb8ac142f64a-Abstract.html.
31. Ikuya Yamada, Akari Asai, and Hannaneh Hajishirzi. Efficient passage retrieval with hashing for open-domain question answering. In *Proceedings of the 59th Annual Meeting of the Association for Computational Linguistics and the 11th International Joint Conference on Natural*

Language Processing (Volume 2: Short Papers), pages 979–986, Online, August 2021. Association for Computational Linguistics. https://doi.org/10.18653/v1/2021.acl-short.123. URL https://aclanthology.org/2021.acl-short.123.

32. Ashish Vaswani, Noam Shazeer, Niki Parmar, Jakob Uszkoreit, Llion Jones, Aidan N. Gomez, Lukasz Kaiser, and Illia Polosukhin. Attention is all you need. In Isabelle Guyon, Ulrike von Luxburg, Samy Bengio, Hanna M. Wallach, Rob Fergus, S. V. N. Vishwanathan, and Roman Garnett, editors, *Advances in Neural Information Processing Systems 30: Annual Conference on Neural Information Processing Systems 2017, December 4–9, 2017, Long Beach, CA, USA*, pages 5998–6008, 2017b. URL https://proceedings.neurips.cc/paper/2017/hash/3f5ee243547dee91fbd053c1c4a845aa-Abstract.html.

33. Xueguang Ma, Kai Sun, Ronak Pradeep, and Jimmy Lin. A replication study of dense passage retriever. *arXiv preprint* arXiv:2104.05740, 2021a.

34. Xilun Chen, Kushal Lakhotia, Barlas Oğuz, Anchit Gupta, Patrick Lewis, Stan Peshterliev, Yashar Mehdad, Sonal Gupta, and Wen-tau Yih. Salient phrase aware dense retrieval: Can a dense retriever imitate a sparse one? *arXiv preprint* arXiv:2110.06918, 2021.

35. Ji Ma, Ivan Korotkov, Yinfei Yang, Keith Hall, and Ryan McDonald. Zero-shot neural passage retrieval via domain-targeted synthetic question generation. *arXiv preprint* arXiv:2004.14503, 2020.

36. Yi Luan, Jacob Eisenstein, Kristina Toutanova, and Michael Collins. Sparse, dense, and attentional representations for text retrieval. *Transactions of the Association for Computational Linguistics*, 9:329–345, 2021.

37. Xueguang Ma, Minghan Li, Kai Sun, Ji Xin, and Jimmy Lin. Simple and effective unsupervised redundancy elimination to compress dense vectors for passage retrieval. In *Proceedings of the 2021 Conference on Empirical Methods in Natural Language Processing*, pages 2854–2859, 2021b.

38. Man Luo, Arindam Mitra, Tejas Gokhale, and Chitta Baral. Improving biomedical information retrieval with neural retrievers. In *Thirty-Sixth AAAI Conference on Artificial Intelligence, AAAI 2022, Thirty-Fourth Conference on Innovative Applications of Artificial Intelligence, IAAI 2022, The Twelveth Symposium on Educational Advances in Artificial Intelligence, EAAI 2022 Virtual Event, February 22 - March 1, 2022*, pages 11038–11046. AAAI Press, 2022a. URL https://ojs.aaai.org/index.php/AAAI/article/view/21352.

39. Tao Chen, Mingyang Zhang, Jing Lu, Michael Bendersky, and Marc Najork. Out-of-domain semantics to the rescue! zero-shot hybrid retrieval models. In *European Conference on Information Retrieval*, pages 95–110. Springer, 2022.

40. Gordon V Cormack, Charles LA Clarke, and Stefan Buettcher. Reciprocal rank fusion outperforms condorcet and individual rank learning methods. In *Proceedings of the 32nd international ACM SIGIR conference on Research and development in information retrieval*, pages 758–759, 2009.

41. Negar Arabzadeh, Xinyi Yan, and Charles LA Clarke. Predicting efficiency/effectiveness trade-offs for dense vs. sparse retrieval strategy selection. In *Proceedings of the 30th ACM International Conference on Information & Knowledge Management*, pages 2862–2866, 2021.

42. Luyu Gao, Zhuyun Dai, Tongfei Chen, Zhen Fan, Benjamin Van Durme, and Jamie Callan. Complement lexical retrieval model with semantic residual embeddings. In *European Conference on Information Retrieval*, pages 146–160. Springer, 2021b.

43. Herve Jegou, Matthijs Douze, and Cordelia Schmid. Product quantization for nearest neighbor search. *IEEE transactions on pattern analysis and machine intelligence*, 33(1):117–128, 2010.

44. Man Luo, Shashank Jain, Anchit Gupta, Arash Einolghozati, Barlas Oguz, Debojeet Chatterjee, Xilun Chen, Chitta Baral, and Peyman Heidari. A study on the efficiency and generalization of light hybrid retrievers. *arXiv preprint* arXiv:2210.01371, 2022b.

45. Tejas Gokhale, Swaroop Mishra, Man Luo, Bhavdeep Singh Sachdeva, and Chitta Baral. Generalized but not robust? comparing the effects of data modification methods on out-of-domain generalization and adversarial robustness. *ArXiv preprint*, abs/2203.07653, 2022a. URL https://arxiv.org/abs/2203.07653.
46. Junnan Li, Dongxu Li, Caiming Xiong, and Steven C. H. Hoi. BLIP: bootstrapping language-image pre-training for unified vision-language understanding and generation. In Kamalika Chaudhuri, Stefanie Jegelka, Le Song, Csaba Szepesvári, Gang Niu, and Sivan Sabato, editors, *International Conference on Machine Learning, ICML 2022, 17–23 July 2022, Baltimore, Maryland, USA*, volume 162 of *Proceedings of Machine Learning Research*, pages 12888–12900. PMLR, 2022b. URL https://proceedings.mlr.press/v162/li22n.html.
47. Alexey Dosovitskiy, Lucas Beyer, Alexander Kolesnikov, Dirk Weissenborn, Xiaohua Zhai, Thomas Unterthiner, Mostafa Dehghani, Matthias Minderer, Georg Heigold, Sylvain Gelly, et al. An image is worth 16x16 words: Transformers for image recognition at scale. *arXiv preprint* arXiv:2010.11929, 2020.
48. Xiyang Dai, Yinpeng Chen, Bin Xiao, Dongdong Chen, Mengchen Liu, Lu Yuan, and Lei Zhang. Dynamic head: Unifying object detection heads with attentions. In *Proceedings of the IEEE/CVF conference on computer vision and pattern recognition*, pages 7373–7382, 2021.
49. Shuai Shao, Zeming Li, Tianyuan Zhang, Chao Peng, Gang Yu, Xiangyu Zhang, Jing Li, and Jian Sun. Objects365: A large-scale, high-quality dataset for object detection. In *Proceedings of the IEEE/CVF international conference on computer vision*, pages 8430–8439, 2019.
50. Aishwarya Kamath, Mannat Singh, Yann LeCun, Gabriel Synnaeve, Ishan Misra, and Nicolas Carion. Mdetr-modulated detection for end-to-end multi-modal understanding. In *Proceedings of the IEEE/CVF International Conference on Computer Vision*, pages 1780–1790, 2021.
51. Piyush Sharma, Nan Ding, Sebastian Goodman, and Radu Soricut. Conceptual captions: A cleaned, hypernymed, image alt-text dataset for automatic image captioning. In *Proceedings of the 56th Annual Meeting of the Association for Computational Linguistics (Volume 1: Long Papers)*, pages 2556–2565, Melbourne, Australia, 2018. Association for Computational Linguistics. https://doi.org/10.18653/v1/P18-1238. URL https://aclanthology.org/P18-1238.
52. Vicente Ordonez, Girish Kulkarni, and Tamara L. Berg. Im2text: Describing images using 1 million captioned photographs. In John Shawe-Taylor, Richard S. Zemel, Peter L. Bartlett, Fernando C. N. Pereira, and Kilian Q. Weinberger, editors, *Advances in Neural Information Processing Systems 24: 25th Annual Conference on Neural Information Processing Systems 2011. Proceedings of a meeting held 12–14 December 2011, Granada, Spain*, pages 1143–1151, 2011. URL https://proceedings.neurips.cc/paper/2011/hash/5dd9db5e033da9c6fb5ba83c7a7ebea9-Abstract.html.
53. Jean-Baptiste Alayrac, Jeff Donahue, Pauline Luc, Antoine Miech, Iain Barr, Yana Hasson, Karel Lenc, Arthur Mensch, Katherine Millican, Malcolm Reynolds, et al. Flamingo: a visual language model for few-shot learning. *Advances in Neural Information Processing Systems*, 35:23716–23736, 2022.
54. Andrew Jaegle, Felix Gimeno, Andy Brock, Oriol Vinyals, Andrew Zisserman, and Joao Carreira. Perceiver: General perception with iterative attention. In *International conference on machine learning*, pages 4651–4664. PMLR, 2021.
55. Kaiming He, Xiangyu Zhang, Shaoqing Ren, and Jian Sun. Deep residual learning for image recognition. In *2016 IEEE Conference on Computer Vision and Pattern Recognition, CVPR 2016, Las Vegas, NV, USA, June 27–30, 2016*, pages 770–778. IEEE Computer Society, 2016. https://doi.org/10.1109/CVPR.2016.90.
56. Jordan Hoffmann, Sebastian Borgeaud, Arthur Mensch, Elena Buchatskaya, Trevor Cai, Eliza Rutherford, Diego de Las Casas, Lisa Anne Hendricks, Johannes Welbl, Aidan Clark, et al. Training compute-optimal large language models. *arXiv preprint* arXiv:2203.15556, 2022.

57. Danny Driess, Fei Xia, Mehdi SM Sajjadi, Corey Lynch, Aakanksha Chowdhery, Brian Ichter, Ayzaan Wahid, Jonathan Tompson, Quan Vuong, Tianhe Yu, et al. Palm-e: An embodied multimodal language model. *arXiv preprint* arXiv:2303.03378, 2023.
58. Muhammad Umer Anwaar, Egor Labintcev, and Martin Kleinsteuber. Compositional learning of image-text query for image retrieval. In *Proceedings of the IEEE/CVF Winter conference on Applications of Computer Vision*, pages 1140–1149, 2021.
59. Man Luo, Yankai Zeng, Pratyay Banerjee, and Chitta Baral. Weakly-supervised visual-retriever-reader for knowledge-based question answering. In *Proceedings of the 2021 Conference on Empirical Methods in Natural Language Processing*, pages 6417–6431, Online and Punta Cana, Dominican Republic, 2021a. Association for Computational Linguistics. https://doi.org/10.18653/v1/2021.emnlp-main.517. URL https://aclanthology.org/2021.emnlp-main.517.
60. Feng Gao, Qing Ping, Govind Thattai, Aishwarya N. Reganti, Ying Nian Wu, and Prem Natarajan. Transform-retrieve-generate: Natural language-centric outside-knowledge visual question answering. In *IEEE/CVF Conference on Computer Vision and Pattern Recognition, CVPR 2022, New Orleans, LA, USA, June 18–24, 2022*, pages 5057–5067. IEEE, 2022a. https://doi.org/10.1109/CVPR52688.2022.00501. URL https://doi.org/10.1109/CVPR52688.2022.00501.
61. Chen Qu, Hamed Zamani, Liu Yang, W Bruce Croft, and Erik Learned-Miller. Passage retrieval for outside-knowledge visual question answering. In *Proceedings of the 44th International ACM SIGIR Conference on Research and Development in Information Retrieval*, pages 1753–1757, 2021.
62. Fabio Petroni, Tim Rocktäschel, Sebastian Riedel, Patrick Lewis, Anton Bakhtin, Yuxiang Wu, and Alexander Miller. Language models as knowledge bases? In *Proceedings of the 2019 Conference on Empirical Methods in Natural Language Processing and the 9th International Joint Conference on Natural Language Processing (EMNLP-IJCNLP)*, pages 2463–2473, Hong Kong, China, 2019a. Association for Computational Linguistics. https://doi.org/10.18653/v1/D19-1250. URL https://aclanthology.org/D19-1250.
63. Liangke Gui, Borui Wang, Qiuyuan Huang, Alexander Hauptmann, Yonatan Bisk, and Jianfeng Gao. KAT: A knowledge augmented transformer for vision-and-language. In *Proceedings of the 2022 Conference of the North American Chapter of the Association for Computational Linguistics: Human Language Technologies*, pages 956–968, Seattle, United States, 2022. Association for Computational Linguistics. https://doi.org/10.18653/v1/2022.naacl-main.70. URL https://aclanthology.org/2022.naacl-main.70.
64. Zhengyuan Yang, Zhe Gan, Jianfeng Wang, Xiaowei Hu, Yumao Lu, Zicheng Liu, and Lijuan Wang. An empirical study of GPT-3 for few-shot knowledge-based VQA. In *Thirty-Sixth AAAI Conference on Artificial Intelligence, AAAI 2022, Thirty-Fourth Conference on Innovative Applications of Artificial Intelligence, IAAI 2022, The Twelveth Symposium on Educational Advances in Artificial Intelligence, EAAI 2022 Virtual Event, February 22 - March 1, 2022*, pages 3081–3089. AAAI Press, 2022. URL https://ojs.aaai.org/index.php/AAAI/article/view/20215.
65. S. Robertson and H. Zaragoza. The probabilistic relevance framework: Bm25 and beyond. *Found. Trends Inf. Retr.*, 3:333–389, 2009.
66. Xiujun Li, Xi Yin, Chunyuan Li, Pengchuan Zhang, Xiaowei Hu, Lei Zhang, Lijuan Wang, Houdong Hu, Li Dong, Furu Wei, et al. Oscar: Object-semantics aligned pre-training for vision-language tasks. In *European Conference on Computer Vision*, pages 121–137. Springer, 2020.
67. Peter Young, Alice Lai, Micah Hodosh, and Julia Hockenmaier. From image descriptions to visual denotations: New similarity metrics for semantic inference over event descriptions. *Transactions of the Association for Computational Linguistics*, 2:67–78, 2014b. https://doi.org/10.1162/tacl_a_00166. URL https://aclanthology.org/Q14-1006.

68. Drew A Hudson and Christopher D Manning. Gqa: a new dataset for compositional question answering over real-world images. *ArXiv preprint*, arXiv:abs/1902.09506, 2019. URL https://arxiv.org/abs/1902.09506.
69. Hao Tan and Mohit Bansal. LXMERT: Learning cross-modality encoder representations from transformers. In *Proceedings of the 2019 Conference on Empirical Methods in Natural Language Processing and the 9th International Joint Conference on Natural Language Processing (EMNLP-IJCNLP)*, pages 5100–5111, Hong Kong, China, 2019b. Association for Computational Linguistics. https://doi.org/10.18653/v1/D19-1514. URL https://aclanthology.org/D19-1514.
70. Patrick S. H. Lewis, Ethan Perez, Aleksandra Piktus, Fabio Petroni, Vladimir Karpukhin, Naman Goyal, Heinrich Küttler, Mike Lewis, Wen-tau Yih, Tim Rocktäschel, Sebastian Riedel, and Douwe Kiela. Retrieval-augmented generation for knowledge-intensive NLP tasks. In Hugo Larochelle, Marc'Aurelio Ranzato, Raia Hadsell, Maria-Florina Balcan, and Hsuan-Tien Lin, editors, *Advances in Neural Information Processing Systems 33: Annual Conference on Neural Information Processing Systems 2020, NeurIPS 2020, December 6–12, 2020, virtual*, 2020a. URL https://proceedings.neurips.cc/paper/2020/hash/6b493230205f780e1bc26945df7481e5-Abstract.html.
71. Feng Gao, Qing Ping, Govind Thattai, Aishwarya Reganti, Ying Nian Wu, and Prem Natarajan. A thousand words are worth more than a picture: Natural language-centric outside-knowledge visual question answering. *ArXiv preprint*, arXiv:abs/2201.05299, 2022b. URL https://arxiv.org/abs/2201.05299.
72. Wonjae Kim, Bokyung Son, and Ildoo Kim. Vilt: Vision-and-language transformer without convolution or region supervision. In Marina Meila and Tong Zhang, editors, *Proceedings of the 38th International Conference on Machine Learning, ICML 2021, 18–24 July 2021, Virtual Event*, volume 139 of *Proceedings of Machine Learning Research*, pages 5583–5594. PMLR, 2021. URL http://proceedings.mlr.press/v139/kim21k.html.
73. Wei-Cheng Chang, Felix X. Yu, Yin-Wen Chang, Yiming Yang, and Sanjiv Kumar. Pre-training tasks for embedding-based large-scale retrieval. In *8th International Conference on Learning Representations, ICLR 2020, Addis Ababa, Ethiopia, April 26–30, 2020*. OpenReview.net, 2020. URL https://openreview.net/forum?id=rkg-mA4FDr.
74. Kenton Lee, Ming-Wei Chang, and Kristina Toutanova. Latent retrieval for weakly supervised open domain question answering. In *Proceedings of the 57th Annual Meeting of the Association for Computational Linguistics*, pages 6086–6096, Florence, Italy, 2019. Association for Computational Linguistics. https://doi.org/10.18653/v1/P19-1612. URL https://aclanthology.org/P19-1612.
75. Kelvin Guu, Kenton Lee, Z. Tung, Panupong Pasupat, and Ming-Wei Chang. Realm: Retrieval-augmented language model pre-training. *ArXiv preprint*, arXiv:abs/2002.08909, 2020. URL https://arxiv.org/abs/2002.08909.
76. Krishna Srinivasan, Karthik Raman, Jiecao Chen, Michael Bendersky, and Marc Najork. Wit: Wikipedia-based image text dataset for multimodal multilingual machine learning. *ArXiv preprint*, arXiv:abs/2103.01913, 2021. URL https://arxiv.org/abs/2103.01913.
77. Yingshan Chang, Mridu Narang, Hisami Suzuki, Guihong Cao, Jianfeng Gao, and Yonatan Bisk. Webqa: Multihop and multimodal qa. In *Proceedings of the IEEE/CVF Conference on Computer Vision and Pattern Recognition*, pages 16495–16504, 2022a.
78. Dhirendra Mohan Datta. *The six ways of knowing: A critical study of the Advaita theory of knowledge*. Motilal Banarsidass, 1997.
79. Shaoqing Ren, Kaiming He, Ross B. Girshick, and Jian Sun. Faster R-CNN: towards real-time object detection with region proposal networks. In Corinna Cortes, Neil D. Lawrence, Daniel D. Lee, Masashi Sugiyama, and Roman Garnett, editors, *Advances in Neural Information Processing Systems 28: Annual Conference on Neural Information Processing Systems 2015, December*

7–12, 2015, Montreal, Quebec, Canada, pages 91–99, 2015. URL https://proceedings.neurips.cc/paper/2015/hash/14bfa6bb14875e45bba028a21ed38046-Abstract.html.
80. Peter Anderson, Xiaodong He, Chris Buehler, Damien Teney, Mark Johnson, Stephen Gould, and Lei Zhang. Bottom-up and top-down attention for image captioning and visual question answering. In *2018 IEEE Conference on Computer Vision and Pattern Recognition, CVPR 2018, Salt Lake City, UT, USA, June 18–22, 2018*, pages 6077–6086. IEEE Computer Society, 2018. https://doi.org/10.1109/CVPR.2018.00636. URL http://openaccess.thecvf.com/content_cvpr_2018/html/Anderson_Bottom-Up_and_Top-Down_CVPR_2018_paper.html.
81. Hao Tan and Mohit Bansal. LXMERT: Learning cross-modality encoder representations from transformers. In *Proceedings of the 2019 Conference on Empirical Methods in Natural Language Processing and the 9th International Joint Conference on Natural Language Processing (EMNLP-IJCNLP)*, pages 5100–5111, Hong Kong, China, 2019a. Association for Computational Linguistics. https://doi.org/10.18653/v1/D19-1514. URL https://aclanthology.org/D19-1514.
82. Tejas Gokhale, Pratyay Banerjee, Chitta Baral, and Yezhou Yang. Vqa-lol: Visual question answering under the lens of logic. In *European conference on computer vision*. Springer, 2020.
83. Tom Kwiatkowski, Jennimaria Palomaki, Olivia Redfield, Michael Collins, Ankur Parikh, Chris Alberti, Danielle Epstein, Illia Polosukhin, Jacob Devlin, Kenton Lee, Kristina Toutanova, Llion Jones, Matthew Kelcey, Ming-Wei Chang, Andrew M. Dai, Jakob Uszkoreit, Quoc Le, and Slav Petrov. Natural questions: A benchmark for question answering research. *Transactions of the Association for Computational Linguistics*, 7:452–466, 2019. https://doi.org/10.1162/tacl_a_00276. URL https://aclanthology.org/Q19-1026.
84. Stanislaw Antol, Aishwarya Agrawal, Jiasen Lu, Margaret Mitchell, Dhruv Batra, C. Lawrence Zitnick, and Devi Parikh. VQA: visual question answering. In *2015 IEEE International Conference on Computer Vision, ICCV 2015, Santiago, Chile, December 7–13, 2015*, pages 2425–2433. IEEE Computer Society, 2015https://doi.org/10.1109/ICCV.2015.279. URL https://doi.org/10.1109/ICCV.2015.279.
85. Yuke Zhu, Oliver Groth, Michael S. Bernstein, and Li Fei-Fei. Visual7w: Grounded question answering in images. In *2016 IEEE Conference on Computer Vision and Pattern Recognition, CVPR 2016, Las Vegas, NV, USA, June 27–30, 2016*, pages 4995–5004. IEEE Computer Society, 2016. https://doi.org/10.1109/CVPR.2016.540. URL https://doi.org/10.1109/CVPR.2016.540.
86. Peng Wang, Qi Wu, Chunhua Shen, Anthony Dick, and Anton Van Den Hengel. Fvqa: Fact-based visual question answering. *IEEE transactions on pattern analysis and machine intelligence*, 40(10):2413–2427, 2017a.
87. Peng Wang, Qi Wu, Chunhua Shen, Anthony R. Dick, and Anton van den Hengel. Explicit knowledge-based reasoning for visual question answering. In Carles Sierra, editor, *Proceedings of the Twenty-Sixth International Joint Conference on Artificial Intelligence, IJCAI 2017, Melbourne, Australia, August 19–25, 2017*, pages 1290–1296. ijcai.org, 2017b. https://doi.org/10.24963/ijcai.2017/179. URL https://doi.org/10.24963/ijcai.2017/179.
88. Sanket Shah, Anand Mishra, Naganand Yadati, and Partha Pratim Talukdar. KVQA: knowledge-aware visual question answering. In *The Thirty-Third AAAI Conference on Artificial Intelligence, AAAI 2019, The Thirty-First Innovative Applications of Artificial Intelligence Conference, IAAI 2019, The Ninth AAAI Symposium on Educational Advances in Artificial Intelligence, EAAI 2019, Honolulu, Hawaii, USA, January 27–February 1, 2019*, pages 8876–8884. AAAI Press, 2019. https://doi.org/10.1609/aaai.v33i01.33018876. URL https://doi.org/10.1609/aaai.v33i01.33018876.
89. Ajeet Kumar Singh, Anand Mishra, Shashank Shekhar, and Anirban Chakraborty. From strings to things: Knowledge-enabled VQA model that can read and reason. In *2019 IEEE/CVF International Conference on Computer Vision, ICCV 2019, Seoul, Korea (South), October 27–*

November 2, 2019, pages 4601–4611. IEEE, 2019. https://doi.org/10.1109/ICCV.2019.00470. URL https://doi.org/10.1109/ICCV.2019.00470.

90. Qingxing Cao, Bailin Li, Xiaodan Liang, Keze Wang, and Liang Lin. Knowledge-routed visual question reasoning: Challenges for deep representation embedding. *IEEE Transactions on Neural Networks and Learning Systems*, 2021.

91. Jie Lei, Licheng Yu, Tamara Berg, and Mohit Bansal. TVQA+: Spatio-temporal grounding for video question answering. In *Proceedings of the 58th Annual Meeting of the Association for Computational Linguistics*, pages 8211–8225, Online, 2020. Association for Computational Linguistics. https://doi.org/10.18653/v1/2020.acl-main.730. URL https://aclanthology.org/2020.acl-main.730.

92. Rowan Zellers, Yonatan Bisk, Ali Farhadi, and Yejin Choi. From recognition to cognition: Visual commonsense reasoning. In *IEEE Conference on Computer Vision and Pattern Recognition, CVPR 2019, Long Beach, CA, USA, June 16–20, 2019*, pages 6720–6731. Computer Vision Foundation / IEEE, 2019. https://doi.org/10.1109/CVPR.2019.00688. URL http://openaccess.thecvf.com/content_CVPR_2019/html/Zellers_From_Recognition_to_Cognition_Visual_Commonsense_Reasoning_CVPR_2019_paper.html.

93. Noa Garcia, Mayu Otani, Chenhui Chu, and Yuta Nakashima. Knowit VQA: answering knowledge-based questions about videos. In *The Thirty-Fourth AAAI Conference on Artificial Intelligence, AAAI 2020, The Thirty-Second Innovative Applications of Artificial Intelligence Conference, IAAI 2020, The Tenth AAAI Symposium on Educational Advances in Artificial Intelligence, EAAI 2020, New York, NY, USA, February 7–12, 2020*, pages 10826–10834. AAAI Press, 2020. URL https://aaai.org/ojs/index.php/AAAI/article/view/6713.

94. Tianran Wu, Noa Garcia, Mayu Otani, Chenhui Chu, Yuta Nakashima, and Haruo Takemura. Transferring domain-agnostic knowledge in video question answering. *ArXiv preprint*, arXiv:abs/2110.13395, 2021b. URL https://arxiv.org/abs/2110.13395.

95. Lisa Anne Hendricks, Subhashini Venugopalan, Marcus Rohrbach, Raymond J. Mooney, Kate Saenko, and Trevor Darrell. Deep compositional captioning: Describing novel object categories without paired training data. In *2016 IEEE Conference on Computer Vision and Pattern Recognition, CVPR 2016, Las Vegas, NV, USA, June 27–30, 2016*, pages 1–10. IEEE Computer Society, 2016. https://doi.org/10.1109/CVPR.2016.8. URL https://doi.org/10.1109/CVPR.2016.8.

96. Di Lu, Spencer Whitehead, Lifu Huang, Heng Ji, and Shih-Fu Chang. Entity-aware image caption generation. In *Proceedings of the 2018 Conference on Empirical Methods in Natural Language Processing*, pages 4013–4023, Brussels, Belgium, 2018. Association for Computational Linguistics. https://doi.org/10.18653/v1/D18-1435. URL https://aclanthology.org/D18-1435.

97. Ali Furkan Biten, Lluís Gómez, Marçal Rusiñol, and Dimosthenis Karatzas. Good news, everyone! context driven entity-aware captioning for news images. In *IEEE Conference on Computer Vision and Pattern Recognition, CVPR 2019, Long Beach, CA, USA, June 16–20, 2019*, pages 12466–12475. Computer Vision Foundation / IEEE, 2019. https://doi.org/10.1109/CVPR.2019.01275. URL http://openaccess.thecvf.com/content_CVPR_2019/html/Biten_Good_News_Everyone_Context_Driven_Entity-Aware_Captioning_for_News_Images_CVPR_2019_paper.html.

98. Alasdair Tran, Alexander Patrick Mathews, and Lexing Xie. Transform and tell: Entity-aware news image captioning. In *2020 IEEE/CVF Conference on Computer Vision and Pattern Recognition, CVPR 2020, Seattle, WA, USA, June 13–19, 2020*, pages 13032–13042. IEEE, 2020. https://doi.org/10.1109/CVPR42600.2020.01305. URL https://doi.org/10.1109/CVPR42600.2020.01305.

99. Spencer Whitehead, Heng Ji, Mohit Bansal, Shih-Fu Chang, and Clare Voss. Incorporating background knowledge into video description generation. In *Proceedings of the 2018 Con-*

ference on Empirical Methods in Natural Language Processing, pages 3992–4001, Brussels, Belgium, 2018. Association for Computational Linguistics. https://doi.org/10.18653/v1/D18-1433. URL https://aclanthology.org/D18-1433.
100. Ting Yao, Yingwei Pan, Yehao Li, and Tao Mei. Incorporating copying mechanism in image captioning for learning novel objects. In *2017 IEEE Conference on Computer Vision and Pattern Recognition, CVPR 2017, Honolulu, HI, USA, July 21–26, 2017*, pages 5263–5271. IEEE Computer Society, 2017. https://doi.org/10.1109/CVPR.2017.559. URL https://doi.org/10.1109/CVPR.2017.559.
101. Yehao Li, Ting Yao, Yingwei Pan, Hongyang Chao, and Tao Mei. Pointing novel objects in image captioning. In *IEEE Conference on Computer Vision and Pattern Recognition, CVPR 2019, Long Beach, CA, USA, June 16–20, 2019*, pages 12497–12506. Computer Vision Foundation / IEEE, 2019a. https://doi.org/10.1109/CVPR.2019.01278. URL http://openaccess.thecvf.com/content_CVPR_2019/html/Li_Pointing_Novel_Objects_in_Image_Captioning_CVPR_2019_paper.html.
102. Niveda Krishnamoorthy, Girish Malkarnenkar, Raymond Mooney, Kate Saenko, and Sergio Guadarrama. Generating natural-language video descriptions using text-mined knowledge. In *Proceedings of the Workshop on Vision and Natural Language Processing*, pages 10–19, Atlanta, Georgia, 2013. Association for Computational Linguistics. URL https://aclanthology.org/W13-1302.
103. Marcella Cornia, Lorenzo Baraldi, Giuseppe Fiameni, and Rita Cucchiara. Universal captioner: Long-tail vision-and-language model training through content-style separation. *ArXiv preprint*, arXiv:abs/2111.12727, 2021. URL https://arxiv.org/abs/2111.12727.
104. Yimin Zhou, Yiwei Sun, and Vasant Honavar. Improving image captioning by leveraging knowledge graphs. In *2019 IEEE winter conference on applications of computer vision (WACV)*, pages 283–293. IEEE, 2019.
105. Feicheng Huang, Zhixin Li, Shengjia Chen, Canlong Zhang, and Huifang Ma. Image captioning with internal and external knowledge. In Mathieu d'Aquin, Stefan Dietze, Claudia Hauff, Edward Curry, and Philippe Cudré-Mauroux, editors, *CIKM '20: The 29th ACM International Conference on Information and Knowledge Management, Virtual Event, Ireland, October 19–23, 2020*, pages 535–544. ACM, 2020. https://doi.org/10.1145/3340531.3411948. URL https://doi.org/10.1145/3340531.3411948.
106. Arushi Goel, Basura Fernando, Thanh-Son Nguyen, and Hakan Bilen. Injecting prior knowledge into image caption generation. In *European Conference on Computer Vision*, pages 369–385. Springer, 2020.
107. Zhiyuan Fang, Tejas Gokhale, Pratyay Banerjee, Chitta Baral, and Yezhou Yang. Video2Commonsense: Generating commonsense descriptions to enrich video captioning. In *Proceedings of the 2020 Conference on Empirical Methods in Natural Language Processing (EMNLP)*, pages 840–860, Online, 2020a. Association for Computational Linguistics. https://doi.org/10.18653/v1/2020.emnlp-main.61. URL https://aclanthology.org/2020.emnlp-main.61.
108. Oleksii Sidorov, Ronghang Hu, Marcus Rohrbach, and Amanpreet Singh. Textcaps: a dataset for image captioning with reading comprehension. In *European Conference on Computer Vision*, pages 742–758. Springer, 2020.
109. Mateusz Malinowski and Mario Fritz. A multi-world approach to question answering about real-world scenes based on uncertain input. In Zoubin Ghahramani, Max Welling, Corinna Cortes, Neil D. Lawrence, and Kilian Q. Weinberger, editors, *Advances in Neural Information Processing Systems 27: Annual Conference on Neural Information Processing Systems 2014, December 8-13 2014, Montreal, Quebec, Canada*, pages 1682–1690, 2014. URL https://proceedings.neurips.cc/paper/2014/hash/d516b13671a4179d9b7b458a6ebdeb92-Abstract.html.

110. Man Luo, Shailaja Keyur Sampat, Riley Tallman, Yankai Zeng, Manuha Vancha, Akarshan Sajja, and Chitta Baral. 'just because you are right, doesn't mean I am wrong': Overcoming a bottleneck in development and evaluation of open-ended VQA tasks. In *Proceedings of the 16th Conference of the European Chapter of the Association for Computational Linguistics: Main Volume*, pages 2766–2771, Online, 2021b. Association for Computational Linguistics. https://doi.org/10.18653/v1/2021.eacl-main.240. URL https://aclanthology.org/2021.eacl-main.240.

Multimodal Content Generation

4

In the previous chapters of the book, we have looked at information retrieval using queries of various formats such as text and visuals. Recently, there has been growing interest both in research and in popular discourse about systems that can generate content from text prompts. These systems which we will call "content generators" allow users to generate content such as images, GIFs, videos, 3D meshes, and audio by simply providing a text prompt. In the context of information retrieval, this new functionality is quickly leading users to a new way to query data-driven systems to retrieve content. The key difference between traditional IR and content generation is that the outputs of IR are retrieved from a known corpus of documents or images, whereas in content generation the outputs are synthesized–i.e. they are artificially generated outputs. Regardless, the growing prowess of content generation has allowed users to use these technologies for many different tasks.

For example, let us consider the task of image retrieval. For this task, the user typically provides a query that reflects the type of information that is being sought. The query may contain, for instance, object categories, actions, descriptions of the background, colors, and so on. Thus the system performing retrieval needs to match such a query with one or more images from an existing database of images. However, recent work in generative adversarial networks [1] and diffusion models [2] for image generation provides us an alternative to searching and returning an image from a database–instead these models can *generate* images from arbitrary text prompts. These generative models are trained on large databases of image-text pairs and once trained can be used to synthesize new images based on the text prompt; these images are likely completely new and "fake", i.e., they were not captured by a camera, but were rather generated using a neural network. On one hand, the fact that the images are fake is problematic as these images may not be accurate and worse–they can be used for malicious purposes to visually portray scenes and events that have never happened in reality [52]. On the other hand, generative modeling allows users to obtain content that the database does not contain–for instance the famous example of an avocado-shaped chair from

[3] which may not have been present in the database but may have been desired by a user for artistic purposes. Generative modeling finds many applications beyond image generation–similar approaches have been used for generating GIFs, videos, audio, 3D meshes, and so on.

4.1 An Anthropological Lens on Visual Content Generation

Content generation isn't exactly a new development–throughout human history, we find evidence of different forms of artwork and media that humans invented to create visual content such as sketches, carvings, drawings, paintings, etc. Digital art is the newest form of media. Generative models that generate content from human-written text are the most recent addition to this long history of visual content generation.

Humans, fascinated by the mystery of our existence, have always looked at things in the observable universe and tried to make sense of those things–the sun, the moon, and the stars; trees, tulips, and hummingbirds; rivers, mountains, and elephants; lakes, meadows, and sand. We perceive the world around us in numerous ways–sight, sound, smell, and touch are the primary sensory inputs that allow us to gain information about the world. Humans, from time immemorial, have looked at the earth and seen things interacting with each other; they have looked at the skies and wondered what the stars and planets were; they have looked at each other and invented communication, cooperation, society, and culture. Humans use that perception not only to make sense of the world but also to interact with the world and (via actions) make changes to the world (Fig. 4.1).

All over the world, we can still find memories of the world as observed and recorded by humans thousands of years ago. For example, this cave painting of hand prints tells us a story of a human community living together 11000 years ago in Argentina. Or this scene that likely depicts collaborative hunting from Egypt, 7000 years ago. These images have allowed our ancestors to communicate what they saw-the environment, other creatures,

Fig. 4.1 Neolithic rock art, over 7,000 years old. Cave of Beasts, Egypt. *Image credits: Clemens Schmillen. License: CC BY-SA 3.0*

4.1 An Anthropological Lens on Visual Content Generation

Fig. 4.2 Some of the earliest known photographs. *On the left* is a photograph taken by John Draper of his sister Dorothy Draper in 1840–this is the earliest surviving photograph of a woman. *In the middle* is the photo titled *"Ruins of Sikandar Bagh"* taken by Felice Beato in 1858 (*Source* the collections of the Imperial War Museums), documenting the aftermath of the First Indian War of Independence in 1857. This is one of the earliest war photographs ever taken, and likely the first-ever photograph of corpses. It is rumored that British occupiers of India or the photographers themselves may have rearranged the skeletal remains of Indian soldiers to heighten the dramatic impact of the photograph. *On the right* is the oldest surviving aerial photograph taken by James Black in 1860 from a hot air balloon floating above Boston Commons

other humans, and their interactions with them. These images, paintings, and various other art forms are human memories in visual form. While it is possible that some of them may not be realistic, and may be exaggerated–even these stories and mythological depictions are human memories in visual form because they represent human experience. In essence, they are a corpus of our shared visual knowledge.

A drawing, painting, or carving is bound to have inaccuracies as it is an approximate and subjective characterization of the scene and is far from perfect. The development of cameras brought about a revolution in the reliable reproduction of scenes. The *camera obscura* as portrayed in Fig. 4.3[1] depicts the earliest attempts at visualizing scenes on flat screens using the pinhole camera model. Pinholes were soon replaced by lenses and took the form of the modern camera.

We have made great strides in developing tools to store images—first on canvas, then on film, and then as 1s and 0s in digital storage. Figure 4.2 shows some of the earliest photographs ever taken–these include portraits, aerial photographs of cities, as well as the dark reality of war.[2, 3, 4] Right from the early phases of photography, images have served as a powerful means of communicating events in our lives. Today, cameras can capture images

---

[1] https://archive.org/details/b30373177/page/n1/mode/2up.
[2] http://westchesterarchives.com/HT/muni/hastings/jwDraperFull.html?catherine.x=51&catherine.y=63.
[3] http://media.iwm.org.uk/iwm/mediaLib/8/media-8388/large.jpg.
[4] https://www.bostonmagazine.com/wp-content/uploads/sites/2/2013/10/tbt.jpg.

Fig. 4.3 This figures shows an illustration of the *camera obscura* (dark chamber). *Source* James Ayscough's *"A short account of the eye and nature of vision"*

in high definition with highly sophisticated acquisition and image processing, and these images can be stored, transmitted, and shared with other people in a fraction of a second over the internet.

In the 2020s, there has been unprecedented growth in the capabilities of models that can generate images from human prompts or descriptions. The growth has been phenomenal in terms of photorealism, i.e. the perceptive quality of such generated images to a point where many of them are indistinguishable from natural images captured by cameras. This new ability to allow humans to type in natural language phrases or sentences (referred to as "prompts") and generate images that correspond to these prompts has been a game-changer in content creation with potential downstream impact in many industries and research fields. Never before has the creation of visual content been so accessible, intuitive, and realistic.

Overview of Chapter. In this chapter, we will review the advances that are being made in this new field of multimodal content generation and also discuss several challenges associated with this emerging technology. First, we will understand the machine learning techniques that drive this technology–most notably, the concept of adversarial learning and diffusion modeling. Then we will learn about how these techniques are applied to several input-to-output mappings, most notably, text-to-image generation, and the current state-of-the-art in image generation under these various input-to-output settings. Finally, we will discuss challenges in the evaluation and benchmarking of various dimensions of multimodal content generation, as well as the risks posed by malicious use of such technology.

4.2 Conditional Image Generation

The aim for conditional image generation models is to generate an image using an input signal. This input signal is typically low-information, i.e. the amount of information contained in the input is lower than the amount of information expected in the output-generated image. Typical examples of low-information inputs are class categories, semantic label maps, sketches, binary masks of objects, point clouds, etc. Compared to all of these inputs, the output is a high-dimensional and therefore, high information signal.

Conditional image generation is then, nothing but the generation of images given some input information about the expected output image; this input will alternatively be called the "condition" in this chapter. This nomenclature is derived from the concept of conditional probability $P(x|y)$ where x is the output and y is the condition. The techniques that seek to learn conditional image generation, essentially estimate the probability distribution $P(x|y)$, thereby allowing the models that estimate the distribution to be able to generate new output images y for a test input x.

Let y be the condition input and x be the target image. Then conditional image generation can be expressed as the function G which maps y to x:

$$x = G(y); \quad G : \mathbb{R}^m \to \mathbb{R}^n, \tag{4.1}$$

where $m < n$ and m, n represent the dimensionality of the input and output respectively.

In the last decade, conditional image generation literature has been dominated by two major approaches: using generative adversarial networks (GANs) [1] and using diffusion models [4].

4.2.1 Conditional GANs

The main difference between conditional and unconditional generation can be understood by the probability distribution that is estimated through the training process. For unconditional generation, a dataset of images is available, and models are trained to estimate the marginal probability distribution $P(x)$ over images x. On the other hand, for conditional image generation, a dataset of images along with *labels* (for instance, class labels for each image or text descriptions for each image) is available, and models are trained to estimate the conditional probability distribution $P(x|y)$ where y are the labels or "conditions" for generation. Mirza and Osindero

Conditional GANs were first proposed by Mirza and Osindero [53] as a novel way to train generative models, compared to *un*conditional image generation. Mirza and Osindero [53] demonstrated their utility by generating images of handwritten digit images conditioned on class labels. cGANs were proposed as a variant of generative adversarial networks that estimated the marginal probability [1].

Generative adversarial networks consist of two networks G and D, denoting a generator and discriminator respectively. The generator G is trained over an image dataset to model the data distribution. The discriminator D is trained to detect which images are fake (i.e. generated by G) and which images belong to the training data. In other words, D estimates the probability of an image coming from the true data distribution. In conditional GANs, both the generator and discriminator are conditioned on additional inputs (the condition y). For the generator, the goal is to combine the condition vector y with a noise vector z sampled from $p_z(z)$ and generate an image. For the discriminator, given y and x, the goal is to discriminate if x is real or fake. G and D are set up in an adversarial game–the generator tries to evade detection by D, while D is trained to improve it's detection capabilities.

The optimization function of a conditional GAN can be expressed as a min-max game between G and D:

$$\min_G \max_D \mathbb{E}_{x \sim p_{data}(x)} log D(x|y) + \mathbb{E}_{z \sim p_z(z)} log(1 - D(G(z|y))). \tag{4.2}$$

Using this min-max function as the objective, model parameters of both G and D are updated until the so-called Nash equilibrium is reached, i.e., from the perspective of the discriminator, the generated image distribution and the real dataset distribution are indistinguishable. In other words, by leveraging a two-player game between the content generator G and critic D, the quality of generated outputs is iteratively improved using typical optimization routines such as gradient descent over the model parameters.

4.2.2 Score-Based Diffusion Models

The main idea behind score-based diffusion models is to learn the score function of the target distribution $p(x)$, where x is an image, using a neural network that produces the score as the score function:

$$s_\theta(x) = \nabla_x \log p(x)$$

The score function measures the likelihood of x being generated by distribution $p(x)$. s_θ is learned via noise-contrastive estimation, which involves adding Gaussian noise to x and training s_θ to approximate the true score function. To generate samples from p(x) using s_θ, we use a diffusion process, which is a Markov chain that starts from a random noise $z_0 \sim N(0, I)$ and gradually transforms it into a sample $x_T \sim p(x)$ by applying small Gaussian perturbations at each step. The diffusion process can be written as:

$$z_{t+1} = \sqrt{1 - \beta_t} z_t + \sqrt{\beta_t} \epsilon_{t+1}, \epsilon_{t+1} \sim N(0, I), t = 0, ..., T - 1$$
$$x_t = z_t + \eta_t, \eta_t \sim N(0, \sigma_t^2 I), t = 0, ..., T$$

where β_t and σ_t^2 are variance reduction and noise level coefficients that control the amount of noise added at each step. The diffusion process preserves the marginal distribution of x_t

at each step, i.e., $x_t \sim p(x)$ for all t; and (2) it has a reverse process that can recover x_0 from x_T using s_θ as follows:

$$z_t = (x_t - \eta_t)/\sqrt{1-\beta_t}, t = 0, ..., T$$
$$\epsilon_t = z_t - \sqrt{1-\beta_t} z_{t-1}, t = 1, ..., T$$
$$\eta_t = -\sigma_t^2 s_\theta(z_t), t = 0, ..., T$$

The reverse process can be seen as a denoising process that removes the noise added by the forward process. By applying the reverse process iteratively from x_T to x_0, we can generate samples from p(x) using $s_\theta(x)$.

To extend score-based diffusion models to conditional image generation, we need to learn the score function of the conditional distribution $p(x|y)$, where y is some condition. There are different ways of estimating the conditional score function using neural networks. One way to do this is via *joint score estimation*, where a single neural network $s_\theta(x, y)$ is trained to produce a score based on both x and y as inputs. Another way is to separate score estimation into two parts: networks $s_\theta(x)$ and $r_\phi(y)$ that takes x and y separately as inputs. The third method is *hybrid score estimation*, which learns a network $s_\theta(x, y)$ that takes both x and y as inputs but also uses $r_\phi(y)$ as an auxiliary network.

4.3 Taxonomy of Conditional Image Generation Tasks

Armed with these two powerful machine learning techniques for conditional image generation, we can now begin to employ them in various tasks that require conditional image generation. Information about images can often be represented in various language-grounded forms–objects present in the image (e.g. *"apple"*), scene categories (e.g. *"a kitchen"*), descriptions of the scene in human language (e.g. *"an apple on a table in a kitchen"*, and so on. Information about images can also be represented in low-dimensional visual forms, such as object bounding boxes, scene layouts, sketches, borders and edges of shapes in the image, semantic segmentation maps, etc. For various reasons including available storage memory, human interaction modalities, and computing workflows, information about an image is often found in one or many of these formats, especially for images on the internet which typically have some metadata associated with them. The goal of conditional image generation, then, is to build models that estimate $P(x|y)$ in order to learn the mapping from metadata or condition space to the space of pixels.

In the subsections that follow, we will explore a few sub-tasks (input-output mappings) under this paradigm of conditional image generation. A pictorial summary of these sub-tasks is presented in Fig. 4.4.

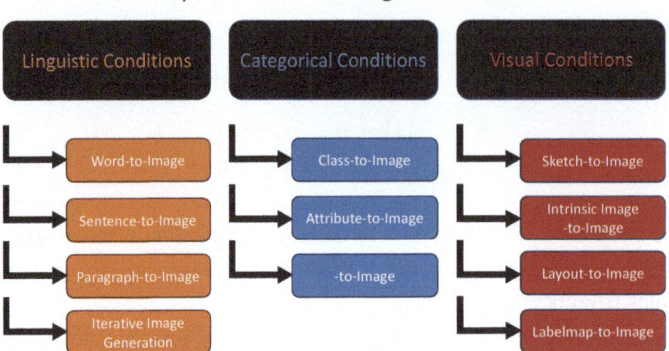

Fig. 4.4 Taxonomy of Conditional Image Generation Tasks. In this chapter, we will review advances in image generation with all three types of conditions mentioned in this figure

4.4 Categorical Conditions for Image Generation

With the proliferation of image classifiers that can detect various categorical representations of images, for example: class labels, attributes (e.g. colors and shapes of objects), latent representations of images, and so on, such categorical information about an image can be easily obtained as predictions from image classifiers. Thus, even when such information is not annotated by humans, the use of image classifiers is the easiest way to obtain categorical conditions, and these conditions can then be used to train image generators.

4.4.1 Category to Image

Although image classifiers may detect one of many semantic categories such as "apple", "dog", "tree", etc. category-to-image models do not use the words associated with these categories. Instead, category-to-image generators use the class labels–usually stored as binary, one-hot, or any other compact latent vectors (e.g. the vector [0,0,1,0,0,0, 0,0,0] represents the digit "2" for a dataset of handwritten digit images, to generate the image. Thus, although the condition might be linked to a word in natural language, such language information is never used to train or use these models.

Mirza and Osindero [53] trained a conditional GAN to generate handwritten digit images given the one-hot class vector as the condition. An example is shown in Fig. 4.5 where each row displays images generated by the conditional GAN, conditioned on digits 0 to 9.

4.4 Categorical Conditions for Image Generation

Fig. 4.5 Outputs of conditional generative adversarial network; each row is conditioned on digits 0 to 9

4.4.2 Image Editing Using Conditional GANs

Image editing models take both a condition and an image as input to generate a novel image. The condition is typically a vector that defines new categories, properties, and attributes that the output image must reflect. For instance, images of birds from the CUB dataset [5] have binary attributes (present/absent) related to sizes, shapes, colors, and other properties of the different parts of bird species. StarGAN and AttGAN [54] are two such models that can edit images conditioned on an attribute vector of those images. These models can be used for the generation of face images conditioned on facial attribute vectors describing the presence or

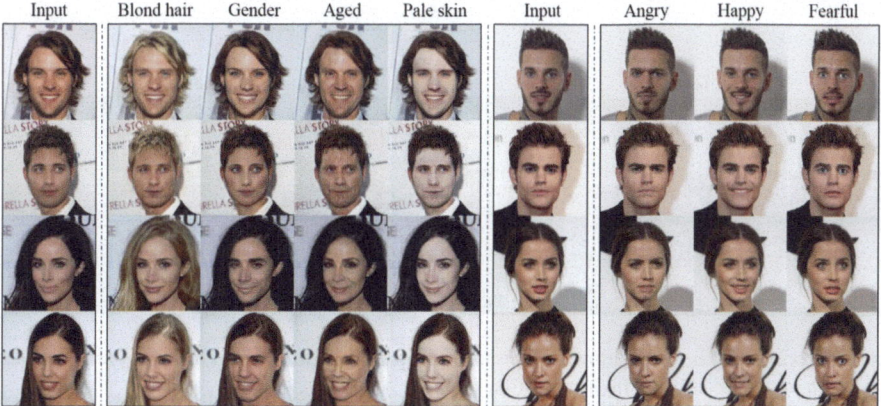

Fig. 4.6 Outputs of StarGAN model that transform the input image by changing the attribute category

absence of facial hairs, glasses, hair color, etc, as shown in Fig. 4.6. Recently [6] used the AttGAN mechanism to perform image augmentations guided by (conditioned on) attributes for improving the generalization of image classifiers.

4.5 Visual Conditions for Image Generation

4.5.1 Semantic Labelmaps to Image

There has been significant research in conditional image generation since cGAN. One such input type used for image generation is a semantic label map. A semantic label map is essentially a matrix of the same size as the image, with each pixel in the matrix corresponding to a category that the pixel belongs to. Figure 4.7 shows an example of a semantic label map on the left and the output of the model on the right. This functionality is one of many image-to-image translation tasks that take as input a condition that is the same size as the image.

Pix2PixHD [7] is a seminal work on semantic label map to image generation, which was demonstrated to synthesize high-resolution and photorealistic outputs using conditional GANs. This work uses multiple conditions such as object instances information in addition to the labelmap. Kulkarni et al. [8] extended this idea in scenarios where the label maps were incomplete, i.e. pixel-wise categorization was missing for most pixels. Recent advances such as SPADE or GauGAN [55] have pushed the state-of-the-art in this sub-task and added more functionalities such as allowing rough human-drawn segmentation maps or even doodles as conditions and generating high-resolution outputs from such approximate label maps as shown in Fig. 4.8 Recent work titled GANCraft [56] has also explored turning semantically labeled block worlds from Minecraft (a popular video game) into photorealistic images as shown in Fig. 4.9.

Fig. 4.7 Pix2PixHD is a conditional generative adversarial network that converts a semantic label map into a photorealistic and high-resolution image

4.5 Visual Conditions for Image Generation

Fig. 4.8 GauGAN allows humans to draw rough hand-drawn label maps and doodles in an interactive user interface. These drawings are converted by GauGAN (a conditional generative model) into photorealistic images

Fig. 4.9 An illustration of GANcraft that can turn semantic label maps from a blocks world domain into photorealistic high-resolution images

4.5.2 Sketch to Image

Humans themselves, are pretty good at generating visual content, so much so, that historical paintings and portraits have often served as a window into historical events. However, *not all of us* possess artistic skills to be able to realistically and accurately depict what we see through such artwork. However, most humans can do a decent job of conveying *salient* aspects of a scene through sketches, some of us better than others. Sketches, often hand-drawn, usually depict salient features of objects such as their shape, boundaries, important features, etc. Humans are very good at understanding the similarities between sketches and photographs of the same concept. For instance, a human-drawn sketch of a cat and a photo of a cat has a large difference in pixel space (i.e. they are visually very different in terms of colors, textures, and other details). However, most humans can deduce the concept (or object type) that a sketch depicts. In some sense, sketches are low-effort and compressed human reproductions of visual concepts. To mimic this human ability to relate sketches to images, sketch-to-image generative models have explored the automatic generation of photorealistic images from hand-drawn scribbles or sketches produced by humans.

Sketch-to-image generation falls into the category of conditional image generation and solutions such as pix2pixHD [7] for other image-to-image tasks such as image generation from semantic labelmaps, can also be applied to sketch-to-image generation. However, one factor that differentiates sketch-to-image generation from these tasks is the lack of pixel-wise alignment of the sketch and the photo–while each pixel of a semantic label map can be assumed to be aligned with the corresponding image pixel, this assumption breaks down for sketches. To overcome these challenges, specialized sketch-to-image generative models have been proposed [9, 57] as well as datasets such as Sketchy [10] and Open-Sketch [11]. A representative work by Koley et al. [12] highlights and defines this problem of pixel-wise alignment in prior work on sketch-to-image generation as the "abstraction gap".

4.6 Text to Image Generation

While many people can sketch and convey the contents of a visual scene, a more universally prevalent human communication method is the use of language to describe a scene. When humans read a sentence, they can often *imagine* and recreate the scene that the sentence describes. Although sentences consider considerably less information than the image, sentences are effective in describing the salient features of an image. Text-to-image (T2I) generation is the challenging task of converting such as low-information input sentence into a dense prediction of pixel intensities. While T2I is a relatively new research area, recent advances have demonstrated exceptional photorealism and intuitive workflows with free-form text prompts. This has resulted in optimism about their utility in commercial products, general use by non-expert users, and also in academic research in computer vision, graphics, and robotics. We are living in an era of unprecedented advances in automated content generation–T2I models have evolved from research prototypes to production-ready tools, with the potential to reshape industries such as entertainment, art, journalism, and education. In this section, we will learn about two major streams of work on text-to-image generation–using generative adversarial networks and diffusion models.

4.6.1 Text-to-Image Generation Using Generative Adversarial Networks

GAN-CLS Generative Adversarial Network with Conditional Latent Space [13] extends the original GAN framework by conditioning both the generator and the discriminator on text embeddings. The text embeddings are obtained by applying a pre-trained text encoder (such as skip-thought vectors) to the input sentences. The generator takes as input a random noise vector and a text embedding and outputs a synthetic image. The discriminator takes as input an image and a text embedding and outputs a probability of whether the image is real or fake, and whether it matches the text or not. The discriminator is trained to classify

4.6 Text to Image Generation

real images that match the text as positive, fake images that match the text as negative, and real images that do not match the text as negative.

The objective function of GAN-CLS consists of two terms: a standard GAN loss that encourages the generator to produce realistic images, and a conditional loss that encourages the generator to produce images that are semantically consistent with the text. The conditional loss is computed by measuring the cosine similarity between the text embeddings and the image features extracted by an auxiliary classifier network.

GAN-CLS was one of the first deep neural networks that was trained end-to-end to synthesize images from input text prompts. However, it was only able to generate low-resolution images on object-specific datasets such as birds, flowers, and faces. This resulted in several limitations in capturing complex details and variations in the text descriptions.

StackGAN Stacked Generative Adversarial Networks) [58] addressed some of the limitations of GAN-CLS by using a two-stage architecture that could generate much larger images of 256 × 256 resolution. The first stage of StackGAN is similar to GAN-CLS: it takes as input a random noise vector and a text embedding and outputs a low-resolution image (64 × 64) that captures the basic shape and color of the object. The second stage of StackGAN takes as input the low-resolution image and a refined text embedding (obtained by applying an attention mechanism to the original text embedding) and outputs a high-resolution image (256 × 256). The objective function of StackGAN consists of three terms: a standard GAN loss for each stage that encourages the generators to produce realistic images, a conditional loss for each stage that encourages the generators to produce images that are semantically consistent with the text, and an additional KL-divergence loss for the second stage that encourages the generators to use diverse noise vectors. While StackGAN was able to capture some complex details and variations in the text descriptions, such as pose, viewpoint, and background, it was limited in its photorealism and was limited to specific datasets. Moreover, it could not handle texts that describe multiple objects or scenes, especially out-of-vocabulary words.

Although many other GAN-based text-to-image variants were proposed in the years following GAN-CLS and StackGAN, it was only after the introduction of diffusion models that we were able to see both photorealism and zero-shot generation from free-form text.

4.6.2 Text-to-Image Generation Using Diffusion Models

From 2016 to 2021, generative adversarial networks (GANs) were the de-facto architectural choice for training text-to-image models. However, in 2021 [59] showed that Diffusion Models could outperform GANs in image synthesis. Results from this work demonstrated superior image quality for the task of unconditional image generation. The work also showed how image quality could be improved in the conditional setting by augmenting diffusion models with classifier guidance–by allowing diffusion models to be conditioned on class labels produced by an image classifier. The classifier is first trained on noisy images and this classifier is used during the diffusion sampling process by using the classifier's gradients to guide the generated images towards the specific class label. Concurrently, another work by Ho and salimans [60] achieved similar results on conditional image generation by a method called "classifier-free guidance" which leveraged interpolations between predictions from a diffusion model with and without labels. Ever since diffusion models have been quickly and widely adopted as a modeling choice for T2I. In this section, we will review the advances that have been made in diffusion models for T2I.

GLIDE. Nichol et al. [14] built on the findings of Ho and salimans [60] and Dhariwal et al. [59] to extend conditional diffusion models to text-to-image synthesis. They explored both approaches: classifier-free guidance and classifier guidance (using CLIP). This approach, called GLIDE, uses a text-encoder with a transformer [15] architecture with 1.2B parameters. This text encoder is trained as part of the image generation training pipeline. GLIDE can be used for text-to-image generation and image inpainting (completion of missing pixels in an image). The paper reports that samples generated by classifier-free guidance are preferred by human evaluators.

Stable Diffusion. Rombach et al. [16] in their seminal work, enabled training diffusion models with limited computational resources, by introducing the new approach of latent diffusion models (LDM). LDMs apply the diffusion process in the latent low-dimensional space of pretrained autoencoders. The training process has two phases. In the first phase, an autoencoder is trained to produce lower-dimensional representations that are perceptually close to the data space. Diffusion models are trained in this latent space which results in higher efficiency. Additionally, the autoencoder only needs to be trained once and can be reused for multiple diffusion models for different image-to-image or text-to-image and other conditioning mechanisms. LDMs are shown to consistently outperform previous GAN-based approaches. This work also showed how latent diffusion models can be used for image super-resolution, image inpainting, class-conditioned image synthesis, and layout-to-image synthesis.

DALL-E. The first version of DALL-E (DALL-E-v1) was released in 2021 [3] for text-to-image generation. DALL-E uses a transformer model that autoregressively models language and visual tokens as a single stream of data. Using pixels directly as image tokens requires high amounts of memory and computational power; to circumvent this, DALL-E uses a two-stage training process. In the first stage, a discrete variational autoencoder is trained to compress images into a 32×32 grid of discrete image tokens where each token can take 8192 possible values. In the second stage, BPE-encoded text tokens are concatenated to the image tokens. An autoregressive transformer is trained to model the joint distribution over these image and text tokens, by maximizing the evidence lower bound [61] on the joint likelihood of the distribution over images, text, and tokens for the encoded image. This approach resulted in substantial improvements in zero-shot text-to-image generation. As shown in Fig. 4.10, the generated outputs are competitive with the other approaches which are trained specifically on MS-COCO data.

4.6 Text to Image Generation

Fig. 4.10 Comparison of outputs generated by DALL-E-v1 with previous GAN-based approaches, tested on the MS-COCO dataset. Figure from [3]

The second version of DALL-E (DALL-E-v2, also known as unCLIP) was released in 2022 [17]. This version leveraged vision-language representations from CLIP [18]. A two-stage modeling strategy is used–in the first stage a prior generates CLIP embeddings from the input text caption and in the second stage a decoder is conditioned on the image embedding to generate the output image. The encoder uses the CLIP image encoder while the decoder is a diffusion model. The advantage of using CLIP embeddings is shown in Fig. 4.11 where images can be smoothly interpolated between two input images or semantically modifying images to move toward the direction of a new text prompt.

The third version of DALL-E (DALL-E-v3) [19] was released in 2023 and showed that text-to-image generation can be substantially improved by training on more descriptive captions. This idea was proposed as a solution to the struggle of existing models to deal with detailed captions and ambiguity because of lack of detail. An image captioning model was developed to re-caption the training dataset by leveraging language models. The main finding was that the re-captioned dataset improved the prompt following ability of T2I models as well as coherence, aesthetics, and compositionality, as well as human preferences.

Imagen and Parti. Imagen [20], is a text-to-image diffusion model which uses generic large language models (pretrained on text-only corpora) as the text encoder and leverages diffusion models for image generation. Imagen uses a large frozen T5-XXL encoder as a text encoder. The embeddings produced by the text encoder are mapped into a 64×64

a photo of a landscape in winter → a photo of a landscape in fall

Fig. 4.11 An illustration of several abilities of DALL-E-v2 beyond text-to-image generation. The first row shows variations between two images by interpolating their CLIP image embeddings and decoding the mixed embedding with a diffusion model. The second row shows the results of interpolating between CLIP image embeddings and a normalized difference between CLIP text embeddings of two different descriptions

image by a conditional diffusion model, and then further upsampled using super-resolution diffusion models to generate high-resolution photorealistic images. Parti [21], short for Pathways Autoregressive Text-to-Image model (Parti), uses an autoregressive model for image generation. Parti formulates the image generation problem as a sequence-to-sequence mapping problem (similar to machine translation or other text-to-text translation problems). This formulation allows the use of large language models in combination with ViT-VQGAN to encode images as sequences of discrete tokens.

4.6.3 Other Variants of Text-to-Image Synthesis

Until this point, we have studied the advances made in architectures, modeling techniques, and training protocols for text-to-image generation that have resulted in the current state of the art in image generation. We will now consider variants of the T2I task where users can interact with T2I models, control them to generate relevant content, iteratively edit generated images, and use T2I models to render images that contain compositions of multiple concepts. The explosion in the abilities of T2I models, powered by diffusion models and efficiency and ease of use due to latent diffusion models has made several exciting applications possible. This phenomenon is already changing the graphics, animation, and media industries.

4.6 Text to Image Generation

Fig. 4.12 An illustration of user-control over attributes of GAN models. Figure adapted from Li et al. [22]

Controllable and Iterative Text-to-Image Generation. T2I models map sentences to pixels, but often cannot account for user preferences, or fine-grained details that need further control from users.

ControlGAN [22] leverages conditional GANs and semantic word embeddings to generate images with user-specified attributes. This is achieved by a mechanism that disentangles different visual attributes and allows the model to manipulate or finetune sub-regions corresponding to the user-specified controls. For example, given a text description of a bird, the user can control the color, shape, and pose of the generated bird by manipulating the word embeddings, as shown in Fig. 4.12.

GeNeVA (Generative Neural Visual Artist) [23] is a model that allows users to iteratively refine the generated image by providing incremental natural language feedback. This approach extends conventional single-step conditional generation into multi-step generation with user feedback. At each time step, the generator generates a new image conditioned on the previous conversation and the image generated in the previous step. Given an initial text description of a scene, an image is generated in the first step–the user can modify this image by adding, removing, or changing objects using natural language commands. GeNeVa uses a variational autoencoder (VAE) to encode both the text and the image into a common latent space, and a decoder to generate the image conditioned on the latent code and the text. SSCR (self-supervised counterfactual reasoning) [62] is another framework that allows iterative language-based image editing. During each time step, the model can edit the image from the previous time step based on new instructions provided by users, as shown in Fig. 4.13.

Promptify [24] is an interactive system that allows users to explore prompts (inputs) that most align with the user's intentions and goals, and allows for refinement of the outputs of text-to-image models to meet that goal. For example, given a text description of a landscape, Promptify can generate images that match the style, mood, and season of the text by using prompts such as "painterly", "sunset", or "winter". Promptify uses a pretrained language model (such as GPT-3) to generate prompts from the text, and a pretrained image model (such as CLIP) to rank the generated images based on the prompts. The interface allows users to organize generated images in order of preference and based on these preferences,

Fig. 4.13 An illustration showing the sequence of images generated by the SSCR model [62]

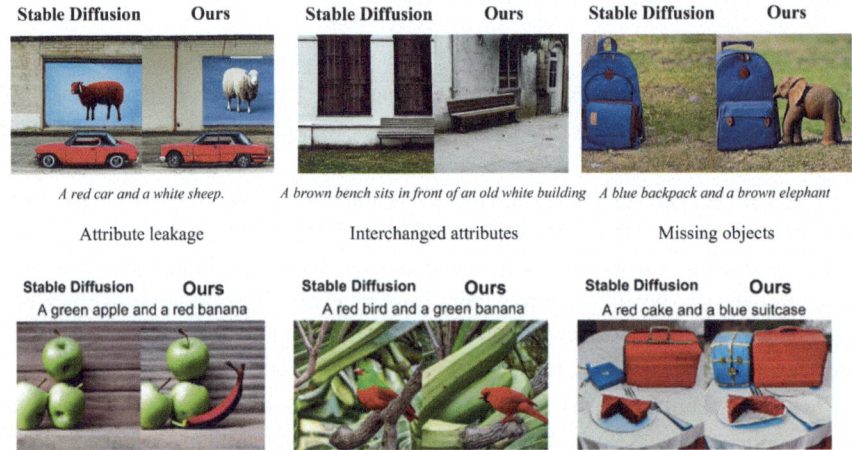

Fig. 4.14 Examples illustrating the improvements in attribute leakage, concept conjugation, compositionality, and multiple objects. Figure adapted from Feng et al. [25]

Promptify suggests potential changes to the original prompt–this iterative feedback enables users to avoid unwanted outcomes and instead reach their desired outcomes.

Semantic Attention Guidance. Feng et al. [25] developed a method dubbed "Structured Diffusion" which incorporates linguistic structure to guide the diffusion process. Language parsers are used to extract spans of text containing visual concepts and are encoded separately to disentangle attribute-object pairs from each other. Results as illustrated by the example in Fig. 4.14 show that this method can help in mitigating attribute leakage, compositionality, and missing objects.

Chefer et al. [26] developed a method that guides T2I models to focus on specific words by intervening in the generative process during inference. The formulation, dubbed "Attend-and-Excite" guides the model to modify the cross-attention values to attend to subject tokens in the input text prompt and strengthen their activations. This process encourages the model to generate all "subjects" described in the text prompt–typically these subjects correspond

4.6 Text to Image Generation

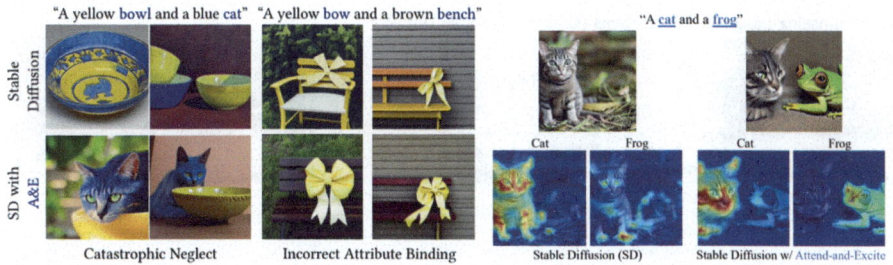

Fig. 4.15 An illustration of the improvements in object generation and attribute binding as a result of Attend-and-Excite (left) and a comparison of the cross-attention maps for subject tokens with and without the Attend-and-Excite method (right). Figure adapted from Chefer et al. [26]

to objects in the image. Examples shown in Fig. 4.15 demonstrate the improvements in generation fidelity when words (highlighted in blue) are stressed during inference.

Composable Diffusion. Liu et al. [27] proposed a method for compositional image generation using diffusion models. This method interprets diffusion models as energy-based models in which data distributions defined by energy functions can be explicitly combined. Two compositional operators: conjunction and negation, are used on top of diffusion models to combine concepts during inference without any additional training or finetuning. This method results in several compositional capabilities as shown in Fig. 4.16.

Image Editing using Stochastic Differential Equations. SDEdit [28] is an image editing and synthesis method based on a diffusion model generative prior, that uses Stochastic Differential Equations (SDE) to iteratively denoise and synthesize photorealistic images. This framework enables several image editing applications such as converting stroke-paintings to images and editing existing images using stroke inputs from users.

4.6.4 Textual Inversion

Text-to-image generative models that we've seen so far are trained to map natural language phrases to images by generative processes such as GANs or diffusion models. Humans describe things that we see in natural language, and we often make use of different concepts such as objects, backgrounds, numbers, relationships, etc. when describing visual scenes. We can therefore view the ability of T2I models to convert descriptions into images as a link between human-level concept understanding and their visual representations. However, it is difficult to introduce novel visual concepts to T2I models, especially those that do not have a very precise language description. Examples are shown in Fig. 4.17. For instance, the example images of the clock on the left all belong to one single type of clock with a certain shape, size, color, etc, The images in the middle are a particular kind of toy cat with stripes and triangular ears, and images on the right are a very specific type of craft. Yet when

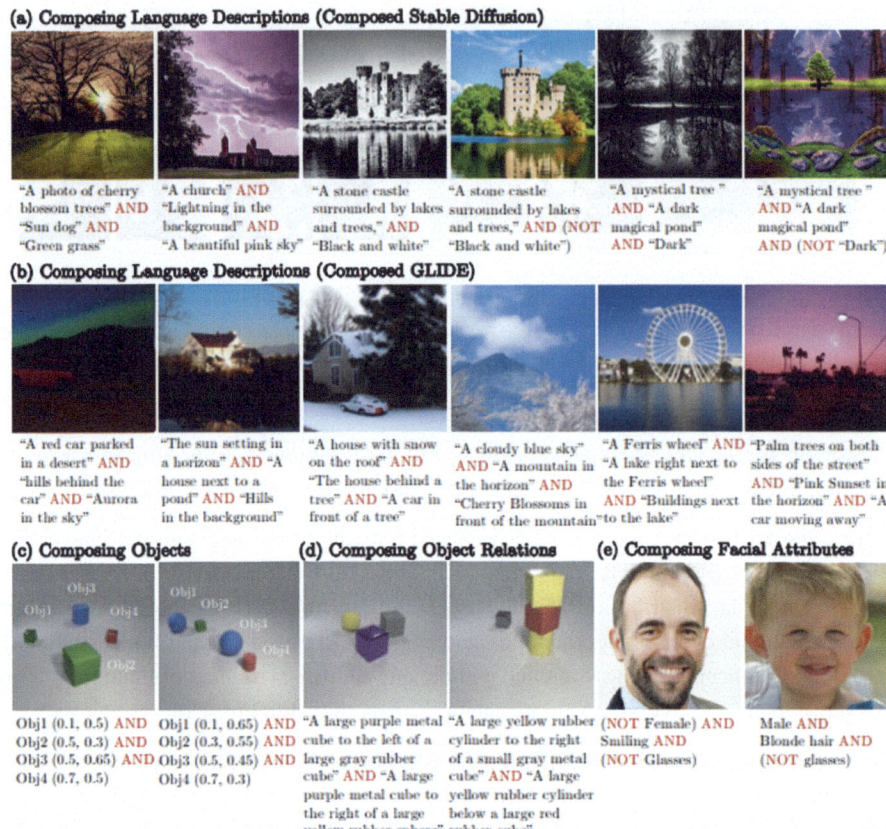

Fig. 4.16 COmposable Diffusion models can generate images from text prompts that are a logical composition of two individual text prompts. The examples in this figure show the images generated for the individual text prompts and the successful generation of their conjunction (with "AND"). Figure from Liu et al. [27]

Fig. 4.17 Examples of novel concepts that are hard to describe with only a few words. Examples from Gal et al. [29]

4.6 Text to Image Generation

humans see images of such novel concepts, we can understand their key features and can relate them to other instances that contain the same concept. Some talented humans (called artists) can also learn these concepts and reproduce them in novel settings in combination with other concepts. In other words, humans can quickly learn concepts, recognize those concepts, and reproduce them compositionally.

Turns out, there are methods that can learn concepts and reproduce them with T2I models, by using a method known as "textual inversion", first proposed by Gal et al. [29]. Textual inversion models operate under the T2I paradigm, but the goal is to use a small number of examples and infer the textual embedding that corresponds to these concepts. The goal is to "find" the textual embedding (in the embedding space of language models). Thus although the users may not have a very precise language description of the images, textual inversion optimizes over the embedding space of language models and finds a representation "S^*" that is shared by these input images.

Let the target concept be c, and let S^* be the text tokens that would represent c., and let us assume access to m images ($X_{1:m}$) of the target concept c. The goal for textual inversion methods is to learn the text tokens S^* corresponding to the concept c from the set of images $X_{1:m}$. This is achieved via an optimization process by reconstructing $X_{1:m}$ using the objective function of the LDM with frozen parameters θ and ϕ:

$$S^* = \operatorname*{argmin}_{s} \mathbb{E}_{\substack{x \in X_{1:m},\, t, \\ \epsilon \sim \mathcal{N}(0,1),\, z \sim \mathcal{E}(x)}} ||\epsilon - \epsilon_\phi(z_t, t, x, C_\theta(y))||_2^2 \qquad (4.3)$$

Once the optimal S^* is obtained, we can generate concept-specific images using the LDM by providing V^* in the text prompt. An example is shown in Fig. 4.18 where four example images of a special kind of sculpture (top) or a special kind of cat toy (bottom) are inverted into a text embedding and reproduced in novel compositions. More approaches along these lines have emerged since the work of Gal et al. [29]. DreamBooth [30] optimizes UNet parameters instead of optimizing the text embedding. Custom Diffusion [31] combines both approaches by optimizing the text embedding S^* and key/value parameters of cross-attention layers. These methods have opened up a new avenue for effectively and quickly learning visual concepts from a few images and reproducing them in novel combinations with other concepts expressed in natural language. The ability of these methods to learn concepts with a very small number of reference images by optimizing for the text embedding that represents the concept or by fine-tuning pre-trained text-conditioned diffusion models is impressive.

4.6.5 Challenges with Evaluation of T2I Models

Over the last 3 years, there has been unprecedented growth in the photorealism of text-to-image generative models. This growth has also reached the common and non-expert user through the democratization of open-source code and several community-built interactive programs that use Stable Diffusion, or proprietary and commercial T2I models such as

Fig. 4.18 An illustration of the textual inversion paradigm, where a very small number of images containing the same complex concept are provided as input. Textual inversion models have the goal of learning this concept as a text embedding "S^*" and reproducing them in novel compositions. Examples from Gal et al. [29]

DALL-E or Midjourney. However this large-scale use of T2I models has started to reveal their limitations, failure modes, and in some cases potential for malicious use. This has led to uncertainty about the capabilities of T2I models in terms of high-fidelity generation that is faithful to the input text prompt.

But how are T2I models evaluated? But when it comes to quantifiable and reproducible evaluation of T2I models, image-level photorealism metrics such as Frechet Inception distance (FID), object accuracy, signal-to-noise ratio, etc. are preferred. In this section, we will review prevalent metrics for evaluating text-to-image generators the limitations of these evaluation metrics, and their ineffectiveness in evaluating complex relationships that exist in images. We will then learn how these limitations are being actively addressed by the research community.

Prevalent T2I Metrics. In 2016, when the first papers on text-to-image generation using neural networks were published [13, 58], the obvious challenge was photorealism, and thus, the performance of T2I models was compared using photorealism metrics such as Inception Score and Frechet Inception Distance. Today, state-of-the-art T2I models have rapidly advanced for users to expect zero-shot generation abilities. Text-to-image synthesis is a relatively new area of research but has seen an explosion in interest and unprecedented improvements in the quality of generated outputs. However, these promising developments have not been accompanied by evaluation protocols that can reliably quantify the performance of T2I models from various perspectives. Gokhale et al. [32] have shown that contemporary metrics are insensitive to errors with generating multiple objects and their relationships, and exhibit several biases, artifacts, and inconsistencies. Similar findings have been reported through small-scale human evaluation studies such as the human study on physical action-based

4.6 Text to Image Generation

Survey of Existing Metrics for T2I Evaluation

	StackGAN (Zhang et al. ICCV 2017)	DM-GAN (Zhu et al. CVPR 2019)	OP-GAN (Hinz et al. TPAMI 2020)	GLIDE (Nichol et al. NeurIPS 2021)	CogView-1/2 (Ding et al. NeurIPS 2021)	DALLE v1/v2 (Ramesh et al. 2021/2022)	Stable Diffusion (Rombach et al. CVPR 2022)
IS: Inception Score (Salimans et al. NeurIPS 2016)	✓	✓	✓	✓	✓	✓	✓
FID: Frechet Inception Distance (Heusel et al. NeurIPS 2017)		✓	✓	✓	✓	✓	✓
R-Precision (Xu et al. CVPR 2018)		✓	✓	✓			
Image Captioning Metrics (Hong et al. CVPR 2018)			✓				
CLIPscore (Hessel et al. EMNLP 2021)				✓			
SOA: Semantic Object Accuracy (Hinz et al. T-PAMI 2020)			✓				
Human Study	✓			✓	✓	✓	✓

Four categories of existing evaluation metrics
1. Purely Visual Metrics for Photorealism : IS, FID
2. Image-text matching : Image Captioning / CLIPscore
3. Object-Level : SOA
4. Human study

Fig. 4.19 A survey of prevalent evaluation metrics and their adoption for quantifying the performance of text-to-image generative models

relations [63], stress-testing of DALLE-v2 [20, 64, 65] in terms of compositionality, grammar, binding, and negation. Developers must evaluate and compare systems using holistic and comprehensive metrics that are a good representation of how humans plan to use text-to-image models. In Fig. 4.19 we overview existing evaluation metrics for T2I and their use by seminal T2I models from 2017 to 2022. These evaluation metrics can be broadly categorized into four categories: purely visual [33, 34], image-text alignment (using captioning [66], retrieval [35] or CLIP [36]), as well as object-based evaluation [67].

Recent Work on Holistic Evaluation In response to the need for evaluating several new dimensions of T2I capabilities, researchers are actively developing new datasets and evaluation frameworks for quantifying the capabilities of T2I models. DALL-Eval [68] proposed several modules for evaluating visual reasoning abilities and also provides a benchmark for evaluating social biases of T2I models(such as gender and skin tone associations with professions and attributes). Spatial reasoning–the ability of T2I models to accurately reproduce simple spatial relationships is the focus of a recent study from Gokhale et al. [32]. This study offers a challenging test dataset (SR2D) and an automated evaluation metric (VISOR) that can quantify the spatial reasoning abilities of T2I models and reveal intriguing biases and artifacts of the models. The ability to perform compositions of different concepts such as objects, actions, attributes, and styles is the focus of a new evaluation tool and dataset called ConceptBed [37]. A new technique: Davidsonian Scene Graphs [38] is an evaluation framework inspired by formal semantics, where an automatic graph-based question generation and question answering module is implemented to produce questions to test images generated by T2I models. The research community is actively working on more evaluation metrics which will eventually contribute to a holistic testing of T2I models. More work needs to be done to develop a comprehensive checklist of model assessments tailored for text-to-image generative models, with emphasis on grounded, automated, and high-fidelity metrics and datasets.

4.7 More Applications of Text-Guided Diffusion Models

Applications of text-guided diffusion modeling are being explored to generate many other types of outputs besides images. This includes short videos and audio clips, as well as three-dimensional scenes such as meshes. We will briefly review some of the latest innovations in this direction.

4.7.1 Text to Video

MuGen (Multimodal Understanding and GENeration). MuGen proposes a framework that can support four tasks: video-audio-text retrieval, video-audio-text generation, video-audio-text captioning, and video-audio-text question answering. Below we summarize each task with examples:

- **Video-Audio-Text Retrieval**: Given a query from one modality (video, audio, or text), retrieve the most relevant items from the other two modalities. For instance, given a video clip of an agent collecting a coin, retrieve the corresponding audio track and text description.
- **Video-Audio-Text Generation**: Given input from one modality (video, audio, or text), generate an output in another modality (video or text) that is consistent with the input. For example, given a text description of an agent talking to an NPC, generate a video clip or an audio track that matches the description.
- **Video-Audio-Text Captioning**: Given input from one or two modalities (video or video-audio), generate a text caption that describes the input. For instance, given a video clip of an agent opening a door, generate a text caption that summarizes what is happening in the video.
- **Video-Audio-Text Question Answering**: Given input from one or two modalities (video or video-audio) and a text question about the input, generate a text answer that answers the question. For example, given a video clip of an agent pushing a box and the question "What color is the box?", generate a text answer that says "The box is green."

Make-a-Video. Make-a-video [39] can automatically generate videos from natural language descriptions by using five modules:

4.7 More Applications of Text-Guided Diffusion Models

(a) A dog wearing a superhero outfit with red cape flying through the sky.

Fig. 4.20 An example of a video generated by MakeAVideo. The input prompt for this example is "a dog wearing a superhero outfit with a red cape flying through the sky". The illustration shows a selection of frames from the video–the motion of the dog can be visualized through these frames

- A natural language parser that extracts the key information from the input text.
- A video retriever that searches for relevant videos from a large database.
- A video composer that aligns the retrieved videos with the input text, using a temporal alignment algorithm and a video summarization method.
- A video synthesizer generates synthetic videos for parts of the text that cannot be matched with existing videos, using a text-to-image model and a video generation model.
- A video editor that merges the retrieved and synthetic videos into a coherent and smooth video, using a voice-over generator, a background music selector, and a transition effect insertion module.

An example is shown in Fig. 4.20.[5]

The paper also evaluates the quality, diversity, and relevance of the generated videos, and concludes by discussing some of the limitations and future directions, such as improving the naturalness and expressiveness of the synthetic videos, incorporating user feedback and preferences, and supporting more languages and domains.

VideoLDM. Video Latent Diffusion Model (VideoLDM) can generate high-resolution videos from text descriptions, by leveraging latent diffusion models (LDMs), which are a type of generative model that can produce high-quality images by training on a compressed latent space. VideoLDM extends LDMs to video generation by introducing a temporal dimension to the latent space and fine-tuning encoded image sequences, i.e., videos. VideoLDM can generate videos with resolutions up to 1280×2048 pixels and 24 frames per second.

VideoLDM has many potential applications in simulation, content creation, and personalization. For example, VideoLDM can generate realistic driving videos for training autonomous vehicles, or create animated videos from text prompts for entertainment purposes. Furthermore, VideoLDM can also perform personalized text-to-video generation, by fine-tuning on different styles or domains of images. This opens up exciting possibilities for future video generation.

---

[5] Note that these are some frames from the video–you can find the complete video examples on the website https://makeavideo.studio/.

Zero Shot Text-to-Video Generation. Text2Video-Zero is a method for generating realistic videos from text descriptions, without using any pre-recorded videos or human annotations. It is a two-stage framework that first generates a semantic layout (containing object location information) from the text and then synthesizes a video from the layout using a generative adversarial network (GAN). The semantic layouts are generated from natural language descriptions using large language models such as GPT3.

All the above work on text-to-video generation has addressed the generation of short (a few seconds) clips and GIFs. However, generating long videos while preserving spatio-temporal context remains a challenge.

4.7.2 Text to Audio

Audio synthesis has been an important research area for many years as generating music, speech, sound effects, etc. finds applications in the movie industry, in the music industry, for video games, and recently for virtual reality applications. While traditional methods for audio synthesis were based on digital signal processing techniques, recent developments in machine learning have also been applied to this field.

Text-to-Audio (TTA) is one step further, giving users the ability to simply type in the description of a scene or an event to generate audio. TTA has recently gained interest due to the success of diffusion models in generating modalities such as images and short videos. It is important to note that TTA is distinct from text-to-speech (TTS). In TTS, a sentence is translated to an audio waveform of speech, i.e. how the sentence would sound like if uttered by a human. In TTA on the other hand, the focus is on generating sound effects that pair well with the input sentence. For instance, given the sentence *"A cat chasing a dog on a beach"*, the goal would be to generate sounds that would match this description–potentially the sound of the ocean, the meowing of a cat, the barking of a dog, and perhaps the sound of their paws hitting the sand.

DiffSound [40] is one such approach that generates sound effects from text prompts. DiffSound contains four modules: a text encoder, a decoder, a VQ-VAE (vector quantized variational autoencoder), and a vocoder. An overview of these modules is shown in Fig. 4.21. The text encoder extracts features from the input text, which are transformed into a mel-spectrogram by the decoder guided by the VQ-VAE, and finally converted into a waveform by the vocoder. The decoder in DiffSound is a non-autoregressive decoder based on the discrete diffusion model. This decoder predicts all mel-spectrogram tokens in a single step and refines them in subsequent steps.

AudioGen [41] tackles the problem of generating audio for descriptive text captions, i.e. sentences that describe visual scenes, such as "a dog barking in the park". This necessitates being able to differentiate between the referred "objects" or separating multiple people speaking simultaneously. AudioGen proposes an augmentation technique to address this challenge. An overview is shown in Fig. 4.22.

4.7 More Applications of Text-Guided Diffusion Models

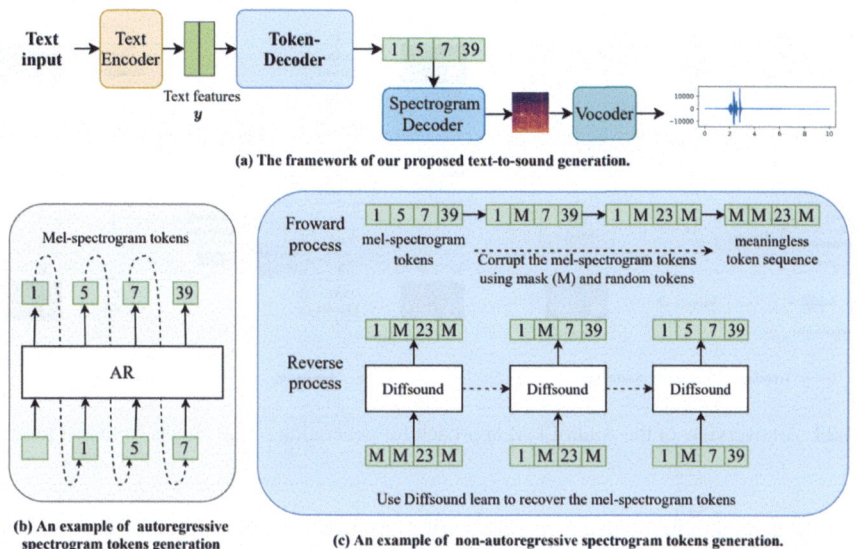

Fig. 4.21 An overview of the DiffSound approach for generating sound effects from text prompts

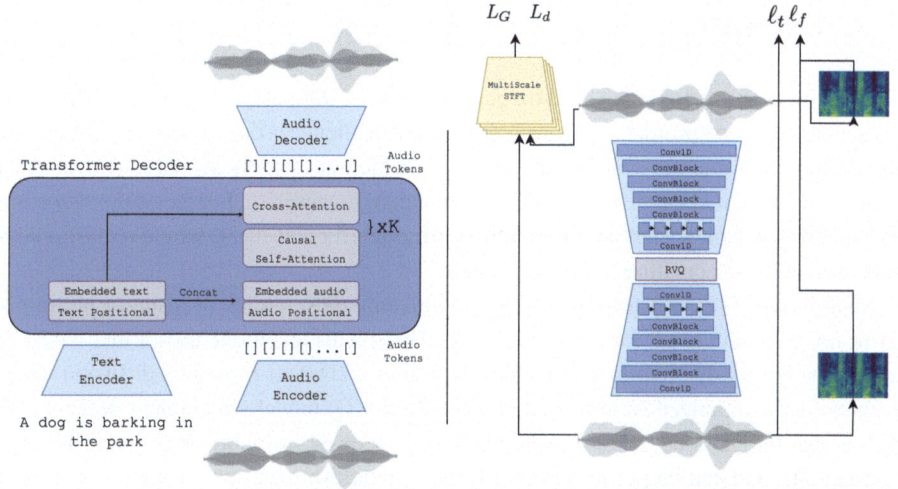

Fig. 4.22 An overview of the AudioGen approach for generating sound effects from text prompts

AudioLDM extends the LDM framework for text-guided audio generation. This framework is designed to generate high-quality audio with long-term consistency. AudioLDM learns continuous audio representations from contrastive language-audio pretraining (CLAP) patients. The pretrained CLAP model allows the LDM to be trained with audio embeddings along with text embeddings as a condition during sampling. AudioLDM also allows text-

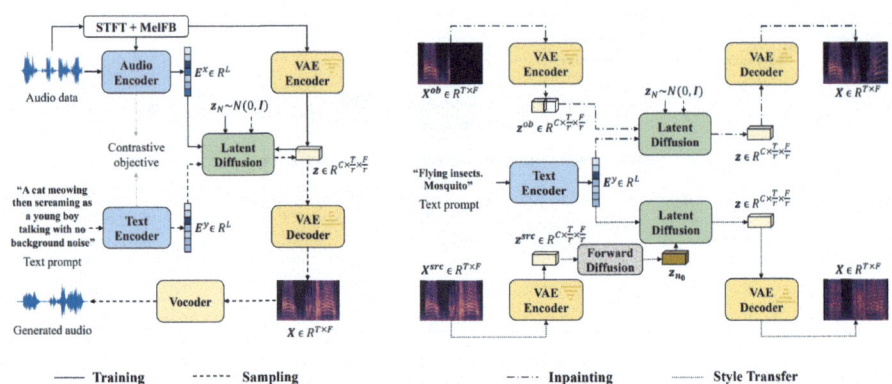

Fig. 4.23 An overview of the AudioLDM approach for generating sound effects from text prompts

guided audio manipulation in a zero-shot setting. An overview of the AudioLDM workflow is shown in Fig. 4.23.

4.7.3 Text to 3D

Diffusion models are also being used for generating three-dimensional content (for instance, 3D meshes) from text prompts. However, the biggest challenge is data–text-to-image models can be trained on internet-scale image-text pairs which are widely available (for instance images and associated captions provided by humans); however such large-scale data is not available for 3D scenes. Thus adopting diffusion models for text-to-3D would require a large annotation effort for creating labeled 3D assets.

DreamFusion [42] is a technique that circumvents these issues by using a pretrained text-to-image model to perform text-to-3D synthesis. This technique uses a loss based on probability density distillation that enables the use of a 2D (text-to-image) diffusion model as a previously optimized Neural Radiance Field (NeRG) model via gradient descent. The result is that the 3D models for a given text can be viewed from any angle, with different illuminations, and can be put in different backgrounds. An overview of the DreamFusion approach is shown in Fig. 4.24.

Magic3D [43] (shown in Fig. 4.25) is a text-to-3D model that can create 3D meshes guided by image conditioning and prompt-based editing. This allows users to control 3D synthesis. A coarse model is obtained first and used as initialization to further optimize a textured 3D mesh model with an efficient differentiable renderer and high-resolution latent diffusion model. This results in a fast rendering time of 40 minutes and higher user preference compared to previous approaches such as DreamFusion.

4.7 More Applications of Text-Guided Diffusion Models

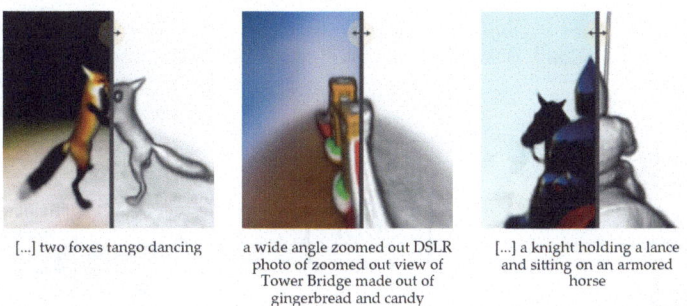

[...] two foxes tango dancing   a wide angle zoomed out DSLR photo of zoomed out view of Tower Bridge made out of gingerbread and candy   [...] a knight holding a lance and sitting on an armored horse

Fig. 4.24 DreamFusion is a technique that uses diffusion models for generating 3D assets from text prompts

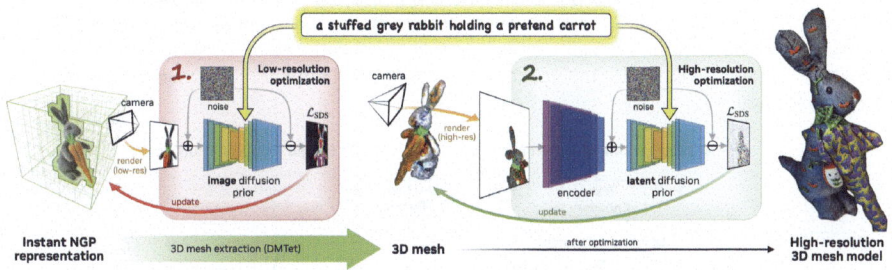

Fig. 4.25 Magic3D creates 3D meshes guided by text prompts

4.7.4 Any-to-Any Generation

In the previous sections, we have seen multiple applications of text-guided diffusion models in generating images, videos, audio, and other modalities. *Any-to-Any Generation* [69] is a recent generative model that uses composable diffusion to jointly generate a combination of video, image, audio, and text from inputs that are also any combinations of video, image, audio, and text. An example is shown in Fig. 4.26 and the architecture is shown in Fig. 4.27. This model, named CoDi (short for Composable Diffusion) first takes an audio track to generate the text output. This is followed by taking an image as an additional output (image+audio) to generate audio. The next step takes image+audio+text as input and generates an image+caption. The model can additionally take image+audio+text to generate video+audio. This new functionality of any-to-any generation could have a significant potential impact on the landscape of multimedia content generation and human-machine interaction in creative domains.

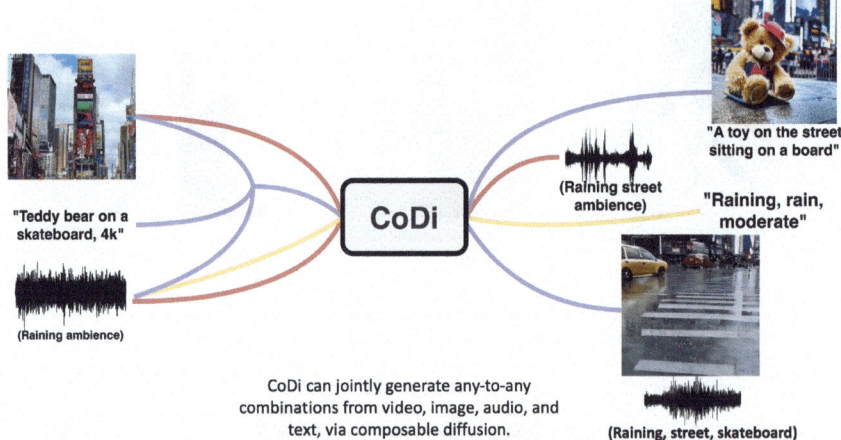

Fig. 4.26 An example of any-to-any generation. Figure from https://codi-gen.github.io/ under CC BY-SA 4.0 license

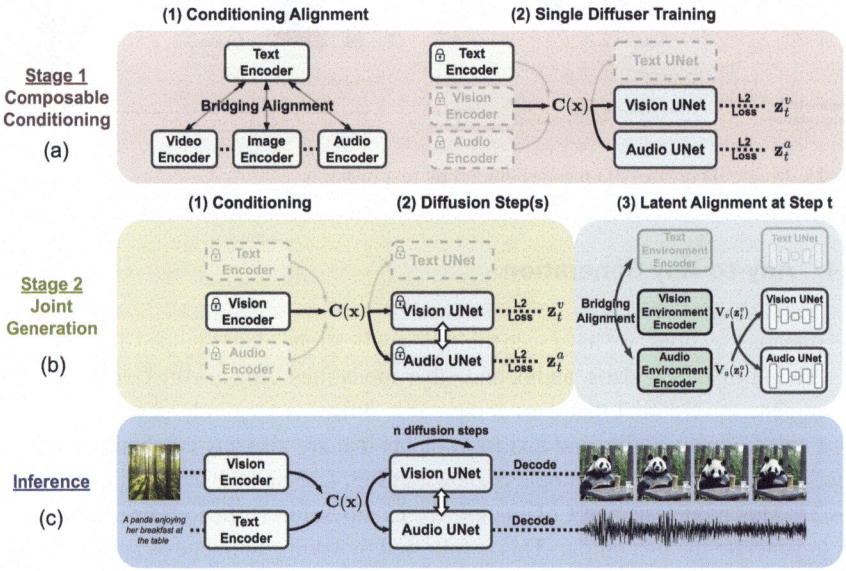

Fig. 4.27 The architecture of the CoDi approach involves three stages: composable conditioning, joint generation, and inference

4.8 Image and Video Captioning

In the previous sections, we discussed generative models which converted input text into visuals. The inverse task–generating natural language descriptions from input images has a much longer history and interest from the research community. The goal of image (or video) captioning systems is to generate a natural language description for input visuals. Image captioning has been an active field of research in the vision community as well as the natural language processing research community. The task of image captioning requires systems to understand images and express that understanding in a format that humans can understand–this is the same format in which humans communicate with each other. For instance, humans often describe the things that they see to each other, in natural language. Thus, image captioning systems are a *communicative* image understanding task, can directly communicate the contents of the image to humans, and have been central to the broader goal of image understanding.

Automated image captioning can be traced back to the work of Gupta et al. [44], Yao et al. [70], Farhadi et al. [45] and Yang et al. [46]. Gupta et al. developed an approach that learned a story-line model of videos using AND-OR graphs and formulated an Integer Programming framework for action recognition and story extraction. This allowed the generation of storylines from new videos. Yao et al. developed a framework to "parse" an image by decomposing input images into visual patterns, converting them into semantic representations using the Web Ontology Language, and a text generation engine that converted those representations into natural language. Both of this approach relied on AND-OR graphs, generated a structured representation of the contents of the scene, and then decoded it into a sentence. Farhadi et al. developed a system to compute a score that links images with sentences. This score can be used for describing images with sentences, as well as obtaining (retrieving) images to match a sentence, by learning a shared "meaning space" onto which both images and sentences can be projected. Yang et al.'s approach predicts nouns, verbs, scenes, and prepositions to make up the core sentence structure and uses a HMM (hidden Markov model) to model the sentence generation process with hidden nodes as sentence components and image detections as the emissions. Early work into image captioning was object- and vocabulary-based–it included detecting actions, detecting objects, attributes, and other scene descriptors, and fitting these detections fitting it into a grammatical structure.

The first approach that used deep neural networks for generating image captioning was proposed by Karpathy and Li [71]. This method leveraged the (at the time) new datasets such as MS-COCO [47] which contained a large set of images and accompanying human-annotated natural language descriptions. The model used a convolutional neural network to process the images and a bidirectional recurrent neural network to model language generation. This method expanded contemporary approaches to model machine translation and language generation using RNNs and expanded that idea conditioned on images for image captioning. Anderson et al. [48] proposed to combine bottom-up and top-down attention mechanisms–this enabled attention to be calculated at the level of objects and salient image regions. This method leveraged advances in faster and better object detection to guide image captioning and visual question answering.

Many subsequent efforts have since used deep neural networks to learn image captioning–these developments have been facilitated by the availability of large-scale datasets which have increased in scale and diversity over the years. Some of the recent image and video captioning approaches are built on the foundation of transformer-based language masked language modeling, extended to the vision domain. These models also leverage web-scale internet data which tends to be noisy. State-of-the-art models such as BLIP [72] utilize such noisy data and produce synthetic captions for web-scale data, paired with a filtering approach to remove noisy captions from the dataset. This approach of bootstrapping has resulted in substantial improvements in various vision-language tasks including image-text retrieval and image captioning under finetuned as well as zero-shot settings.

4.8.1 Human-in-the-Loop Image Captioning

Image captioning systems that we saw above are typically formulated as neural networks convert input images x into captions y through a generative process $y = g(x)$. These models are trained on image captioning datasets such as MS-COCO [49] which contains images and human-annotated captions corresponding to those images. As such, a trained caption generator $g()$ learns the mapping between the inputs and outputs in the dataset but is often not creative. More importantly, humans cannot directly control or influence the generative of these captions, and there is no opportunity for interactive editing of these captions. Automatically generated captions often do not contain satisfactory content as images contain several objects, backgrounds, and contexts that cannot be covered in a single sentence.

To address these issues, human-in-the-loop or interactive captioning has been recently explored. Zheng et al. [73] introduced a framework for human-in-the-loop captioning–as part of this framework, humans can provide a set of keywords as an additional input as a condition to guide caption generation. This additional input acts as a weak supervision for the model to learn from the user and improve, enhance, and edit the caption. Padmakumar and He [74] introduced an approach to re-write captions through human inputs. Through user studies, they found that machine-written captions often fall short of satisfying user intentions and that human users prefer captions generated through a collaborative process between humans and caption generators. An overview of this approach is shown in Fig. 4.29.

4.8 Image and Video Captioning

Fig. 4.28 Image captioning systems describe images in human languages such as English, often describing the objects present in the images, and sometimes describing the relationships between those objects, and the relationship between the objects in the context of the scene. For example, a typical image captioning system might generate a sentence such as *"Two people in boats in a river"*

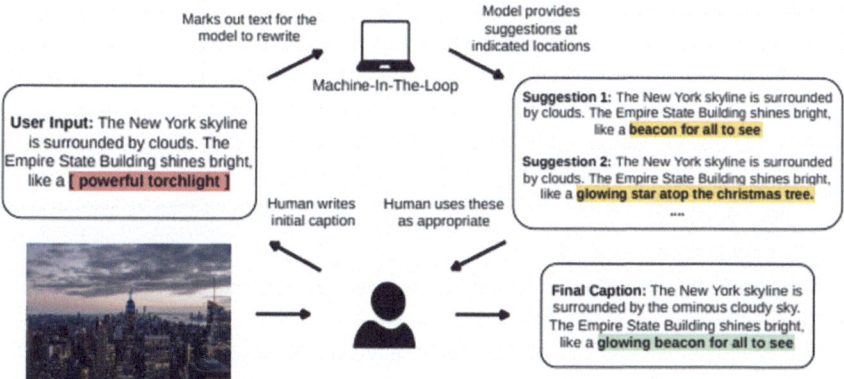

Fig. 4.29 An overview of machine-in-the-loop caption rewriting. Figure from Padmakumar and He [74]

4.8.2 Image Captioning with Commonsense Reasoning

Conventional image captioning systems that we have seen in the first part of this section typically generate sentences that describe objects, actions, and the overall background or "situation". For example, in Fig. 4.28, a state-of-the-art image captioning system such as

BLIP [72] that uses vision and language transformers[6] generates captions such as:

> "two people in yellow boats paddle through a body of water"
> "two people kayaking in the water with a city in the background"
> "two people kayaking on a lake with the city skyline in the background"
> "two young men in kayaks on a river with a city behind them"

Do these sentences describe these images *entirely*? Obviously not. There are many *other* things in the image and there is a lot of context and background information or prior knowledge of the scene that hasn't been described in the caption. This is often the case with image captioning systems–a sentence is a low-information compressed representation of the image.

Which aspects of the image haven't been described in the above captions?

For starters, specifics are missing! What city is it? Answering that would require reasoning with knowledge about scenes in Pittsburgh (which happens to be the city that this picture was taken in)–dedicated Pittsburghers would identify that these two people are kayaking at the confluence of the two rivers Monongahela and Allegheny and that the buildings behind them are in Downtown Pittsburgh, and that there is a tiny fountain in the background and that is Point State Park, and on and on. But this knowledge is also available in data sources (either image-based or text-based corpora) but none of the above information is *in the image*.

Humans often reason about scenes in speculative ways. For example, is the person on the right holding the paddle correctly? Does that indicate that he may not be an expert kayaker? What could happen if he loses the paddle? Humans also inherently infer attributes about people performing actions (for example, whether or not the person on the right is a professional kayaker), the intentions behind people's actions, and the effects of those actions on the world. As such, the development of image captioning systems that can incorporate such commonsense reasoning or speculative descriptions of the scene is a new research direction being pursued.

VisualComet [50] contains an image captioning dataset that contains commonsense annotations for events that occur before and after an image frame, as shown in Fig. 4.30. In Video2Commonsense [51], a video captioning dataset was developed that allowed models to go beyond conventional descriptive captions–these caption generators were trained to generate intentions of people's actions and the effects of those actions, along with some attributes of the people performing actions. The example in Fig. 4.31 shows the output of a conventional video captioning system: *"group of runners get prepared to run a race"*. By training models on the Video2Commonsense dataset, this caption is enhanced by intentions of the agents: *"to win a medal"*, the effect of the agents' actions: *"they are congratulated at the finish line"*, and an attribute of the agents: *"athletic"*. The Video2Commonsense dataset also contains annotated question-answer pairs to reason about the intentions of agents, their effects, and attributes. Note that the predictions made by the models are speculative, as there

---

[6] We used the demo from https://huggingface.co/spaces/Salesforce/BLIP.

4.9 Broader Impact of Multimodal Content Generation

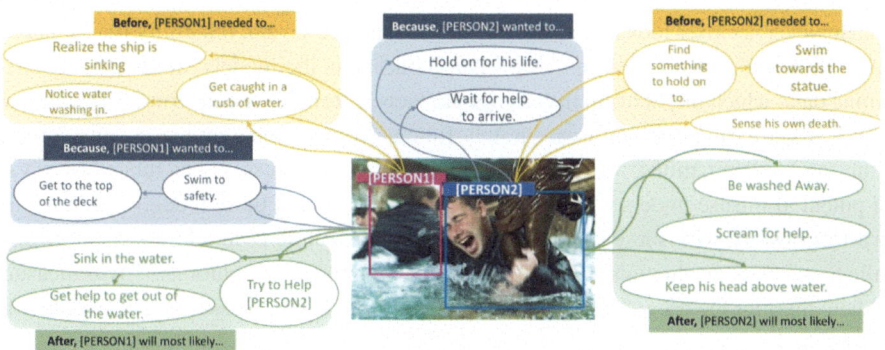

Fig. 4.30 An example from the Visual Comet dataset showing the task of describing preconditions, intentions, and effects of actions performed by people

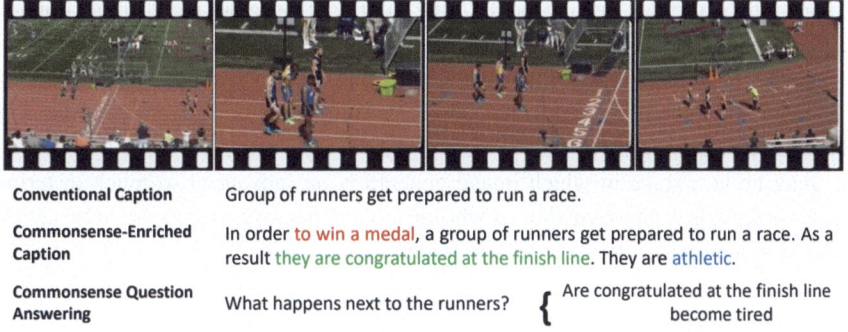

Fig. 4.31 An example illustrating the difference between traditional video captions that describe agents and their actions and commonsense-enriched captions that augment the caption with speculation about intentions and effects of actions and abilities or attributes of the agents performing them

often isn't a way to evaluate the correctness of the intentions of agents performing actions from videos. However, these systems allow the captioning model to become more creative and generate captions in a way that reflects human preferences and interpretations.

4.9 Broader Impact of Multimodal Content Generation

The surprising efficacy, ease of use, and widespread access of multimodal content generators, especially text-to-image diffusion models, have several implications. Over the past few years, generative models have evolved from simple research concepts to production-ready tools, dramatically reshaping the tech landscape. Their outstanding generative capabilities have gained traction in various sectors, such as entertainment, art, journalism, and education. However, a closer look reveals that these models face several reliability issues that can impact their widespread adoption. A primary concern is the models' ability to memorize training data, which might result in copyright breaches. Reliability concerns also encompass the model's occasional failure to accurately follow prompts, inherent biases, misrepresentations, and hallucinations. Moreover, with increasing awareness, issues related to privacy and potential misuse underscore the urgent need to safeguard these models. To move forward responsibly with these models, we must adopt solutions to address memorization challenges, robust evaluation systems, and active fingerprinting solutions. These measures will help monitor the progress and ensure responsible and effective use of image-generative models.

For many image retrieval applications in which exact matches may not be necessary and approximately correct images may be satisfactory, content generation algorithms could be employed instead. Consider the case of image retrieval, for instance, for finding stock images to demonstrate two people shaking hands. If the identity of those two people is not important, and any image of two people shaking hands would suffice, text-to-image generative models can be deployed with the input prompt *"two people shaking hands"* and images can be generated within seconds, as opposed to a time-consuming search and retrieval from a large database of stock photographs.

We may be at a stage in which image generation has advanced so much in terms of photorealism, fidelity, and zero-shot capabilities that generative models could be seriously considered as alternatives to retrieval algorithms.

References

1. Ian J. Goodfellow, Jean Pouget-Abadie, Mehdi Mirza, Bing Xu, David Warde-Farley, Sherjil Ozair, Aaron C. Courville, and Yoshua Bengio. Generative adversarial nets. In Zoubin Ghahramani, Max Welling, Corinna Cortes, Neil D. Lawrence, and Kilian Q. Weinberger, editors, *Advances in Neural Information Processing Systems 27: Annual Conference on Neural Information Processing Systems 2014, December 8–13 2014, Montreal, Quebec, Canada*, pages 2672–2680, 2014. URL https://proceedings.neurips.cc/paper/2014/hash/5ca3e9b122f61f8f06494c97b1afccf3-Abstract.html.
2. Jascha Sohl-Dickstein, Eric A. Weiss, Niru Maheswaranathan, and Surya Ganguli. Deep unsupervised learning using nonequilibrium thermodynamics. In Francis R. Bach and David M. Blei, editors, *Proceedings of the 32nd International Conference on Machine Learning, ICML 2015, Lille, France, 6–11 July 2015*, volume 37 of *JMLR Workshop and Conference Proceedings*, pages 2256–2265. JMLR.org, 2015. URL http://proceedings.mlr.press/v37/sohl-dickstein15.html.
3. Aditya Ramesh, Mikhail Pavlov, Gabriel Goh, Scott Gray, Chelsea Voss, Alec Radford, Mark Chen, and Ilya Sutskever. Zero-shot text-to-image generation. In Marina Meila and Tong Zhang, editors, *Proceedings of the 38th International Conference on Machine Learning, ICML 2021, 18–24 July 2021, Virtual Event*, volume 139 of *Proceedings of Machine Learning Research*, pages 8821–8831. PMLR, 2021. URL http://proceedings.mlr.press/v139/ramesh21a.html.
4. Yang Song, Jascha Sohl-Dickstein, Diederik P Kingma, Abhishek Kumar, Stefano Ermon, and Ben Poole. Score-based generative modeling through stochastic differential equations. In *Inter-

national Conference on Learning Representations, 2020.
5. Peter Welinder, Steve Branson, Takeshi Mita, Catherine Wah, Florian Schroff, Serge Belongie, and Pietro Perona. Caltech-ucsd birds 200. Technical Report CNS-TR-201, Caltech, 2010. URL http://www.se3/wp-content/uploads/2014/09/WelinderEtal10_CUB-200.pdf, http://www.vision.caltech.edu/visipedia/CUB-200.html.
6. Tejas Gokhale, Rushil Anirudh, Bhavya Kailkhura, Jayaraman J. Thiagarajan, Chitta Baral, and Yezhou Yang. Attribute-guided adversarial training for robustness to natural perturbations. In *Thirty-Fifth AAAI Conference on Artificial Intelligence, AAAI 2021, Thirty-Third Conference on Innovative Applications of Artificial Intelligence, IAAI 2021, The Eleventh Symposium on Educational Advances in Artificial Intelligence, EAAI 2021, Virtual Event, February 2–9, 2021*, pages 7574–7582. AAAI Press, 2021. URL https://ojs.aaai.org/index.php/AAAI/article/view/16927.
7. Ting-Chun Wang, Ming-Yu Liu, Jun-Yan Zhu, Andrew Tao, Jan Kautz, and Bryan Catanzaro. High-resolution image synthesis and semantic manipulation with conditional gans. In *2018 IEEE Conference on Computer Vision and Pattern Recognition, CVPR 2018, Salt Lake City, UT, USA, June 18–22, 2018*, pages 8798–8807. IEEE Computer Society, 2018. https://doi.org/10.1109/CVPR.2018.00917. URL http://openaccess.thecvf.com/content_cvpr_2018/html/Wang_High-Resolution_Image_Synthesis_CVPR_2018_paper.html.
8. Kuldeep Kulkarni, Tejas Gokhale, Rajhans Singh, Pavan Turaga, and Aswin Sankaranarayanan. Halluci-net: Scene completion by exploiting object co-occurrence relationships. *ArXiv preprint*, abs/2004.08614, 2020. URL arXiv:2004.08614.
9. Cusuh Ham, Gemma Canet Tarres, Tu Bui, James Hays, Zhe Lin, and John Collomosse. Cogs: Controllable generation and search from sketch and style. *European Conference on Computer Vision*, 2022.
10. Patsorn Sangkloy, Nathan Burnell, Cusuh Ham, and James Hays. The sketchy database: learning to retrieve badly drawn bunnies. *ACM Transactions on Graphics (TOG)*, 35(4): 1–12, 2016.
11. Yulia Gryaditskaya, Mark Sypesteyn, Jan Willem Hoftijzer, Sylvia Pont, Frédo Durand, and Adrien Bousseau. Opensketch: a richly-annotated dataset of product design sketches. *ACM Transactions on Graphics (TOG)*, 38(6): 1–16, 2019.
12. Subhadeep Koley, Ayan Kumar Bhunia, Aneeshan Sain, Pinaki Nath Chowdhury, Tao Xiang, and Yi-Zhe Song. Picture that sketch: Photorealistic image generation from abstract sketches. In *Proceedings of the IEEE/CVF Conference on Computer Vision and Pattern Recognition*, pages 6850–6861, 2023.
13. Scott E. Reed, Zeynep Akata, Xinchen Yan, Lajanugen Logeswaran, Bernt Schiele, and Honglak Lee. Generative adversarial text to image synthesis. In Maria-Florina Balcan and Kilian Q. Weinberger, editors, *Proceedings of the 33nd International Conference on Machine Learning, ICML 2016, New York City, NY, USA, June 19–24, 2016*, volume 48 of *JMLR Workshop and Conference Proceedings*, pages 1060–1069. JMLR.org, 2016. URL http://proceedings.mlr.press/v48/reed16.html.
14. Alexander Quinn Nichol, Prafulla Dhariwal, Aditya Ramesh, Pranav Shyam, Pamela Mishkin, Bob McGrew, Ilya Sutskever, and Mark Chen. GLIDE: towards photorealistic image generation and editing with text-guided diffusion models. In Kamalika Chaudhuri, Stefanie Jegelka, Le Song, Csaba Szepesvári, Gang Niu, and Sivan Sabato, editors, *International Conference on Machine Learning, ICML 2022, 17–23 July 2022, Baltimore, Maryland, USA*, volume 162 of *Proceedings of Machine Learning Research*, pages 16784–16804. PMLR, 2022. URL https://proceedings.mlr.press/v162/nichol22a.html.
15. Ashish Vaswani, Noam Shazeer, Niki Parmar, Jakob Uszkoreit, Llion Jones, Aidan N. Gomez, Lukasz Kaiser, and Illia Polosukhin. Attention is all you need. In Isabelle Guyon, Ulrike von Luxburg, Samy Bengio, Hanna M. Wallach, Rob Fergus, S. V. N. Vishwanathan, and

Roman Garnett, editors, *Advances in Neural Information Processing Systems 30: Annual Conference on Neural Information Processing Systems 2017, December 4–9, 2017, Long Beach, CA, USA*, pages 5998–6008, 2017b. URL https://proceedings.neurips.cc/paper/2017/hash/3f5ee243547dee91fbd053c1c4a845aa-Abstract.html.

16. Robin Rombach, Andreas Blattmann, Dominik Lorenz, Patrick Esser, and Björn Ommer. High-resolution image synthesis with latent diffusion models. In *Proceedings of the IEEE/CVF Conference on Computer Vision and Pattern Recognition*, pages 10684–10695, 2022.

17. Aditya Ramesh, Prafulla Dhariwal, Alex Nichol, Casey Chu, and Mark Chen. Hierarchical text-conditional image generation with clip latents. *ArXiv preprint*, abs/2204.06125, 2022. URL arXiv:2204.06125.

18. Alec Radford, Jong Wook Kim, Chris Hallacy, Aditya Ramesh, Gabriel Goh, Sandhini Agarwal, Girish Sastry, Amanda Askell, Pamela Mishkin, Jack Clark, Gretchen Krueger, and Ilya Sutskever. Learning transferable visual models from natural language supervision. In Marina Meila and Tong Zhang, editors, *Proceedings of the 38th International Conference on Machine Learning, ICML 2021, 18–24 July 2021, Virtual Event*, volume 139 of *Proceedings of Machine Learning Research*, pages 8748–8763. PMLR, 2021. URL http://proceedings.mlr.press/v139/radford21a.html.

19. James Betker, Gabriel Goh, Li Jing, Tim Brooks, Jianfeng Wang, Linjie Li, Long Ouyang, Juntang Zhuang, Joyce Lee, Yufei Guo, et al. Improving image generation with better captions, 2023.

20. Chitwan Saharia, William Chan, Saurabh Saxena, Lala Li, Jay Whang, Emily L Denton, Kamyar Ghasemipour, Raphael Gontijo Lopes, Burcu Karagol Ayan, Tim Salimans, et al. Photorealistic text-to-image diffusion models with deep language understanding. *Advances in Neural Information Processing Systems*, 35: 36479–36494, 2022.

21. Jiahui Yu, Yuanzhong Xu, Jing Yu Koh, Thang Luong, Gunjan Baid, Zirui Wang, Vijay Vasudevan, Alexander Ku, Yinfei Yang, Burcu Karagol Ayan, Ben Hutchinson, Wei Han, Zarana Parekh, Xin Li, Han Zhang, Jason Baldridge, and Yonghui Wu. Scaling autoregressive models for content-rich text-to-image generation. *Transactions on Machine Learning Research*, 2022. ISSN 2835-8856. URL https://openreview.net/forum?id=AFDcYJKhND. Featured Certification.

22. Bowen Li, Xiaojuan Qi, Thomas Lukasiewicz, and Philip H. S. Torr. Controllable text-to-image generation. In Hanna M. Wallach, Hugo Larochelle, Alina Beygelzimer, Florence d'Alché-Buc, Emily B. Fox, and Roman Garnett, editors, *Advances in Neural Information Processing Systems 32: Annual Conference on Neural Information Processing Systems 2019, NeurIPS 2019, December 8–14, 2019, Vancouver, BC, Canada*, pages 2063–2073, 2019b. URL https://proceedings.neurips.cc/paper/2019/hash/1d72310edc006dadf2190caad5802983-Abstract.html.

23. Alaaeldin El-Nouby, Shikhar Sharma, Hannes Schulz, Devon Hjelm, Layla El Asri, Samira Ebrahimi Kahou, Yoshua Bengio, and Graham W Taylor. Keep drawing it: Iterative language-based image generation and editing. In *Neural Information Processing Systems: Visually Grounded Interaction and Language Workshop*, 2018.

24. Stephen Brade, Bryan Wang, Mauricio Sousa, Sageev Oore, and Tovi Grossman. Promptify: Text-to-image generation through interactive prompt exploration with large language models. In *Proceedings of the 36th Annual ACM Symposium on User Interface Software and Technology*, pages 1–14, 2023.

25. Weixi Feng, Xuehai He, Tsu-Jui Fu, Varun Jampani, Arjun Reddy Akula, Pradyumna Narayana, Sugato Basu, Xin Eric Wang, and William Yang Wang. Training-free structured diffusion guidance for compositional text-to-image synthesis. In *The Eleventh International Conference on Learning Representations*, 2023. URL https://openreview.net/forum?id=PUIqjT4rzq7.

26. Hila Chefer, Yuval Alaluf, Yael Vinker, Lior Wolf, and Daniel Cohen-Or. Attend-and-excite: Attention-based semantic guidance for text-to-image diffusion models, 2023.

27. Nan Liu, Shuang Li, Yilun Du, Antonio Torralba, and Joshua B Tenenbaum. Compositional visual generation with composable diffusion models. *European Conference on Computer Vision*, 2022.
28. Chenlin Meng, Yutong He, Yang Song, Jiaming Song, Jiajun Wu, Jun-Yan Zhu, and Stefano Ermon. Sdedit: Guided image synthesis and editing with stochastic differential equations. In *International Conference on Learning Representations*, 2021.
29. Rinon Gal, Yuval Alaluf, Yuval Atzmon, Or Patashnik, Amit Haim Bermano, Gal Chechik, and Daniel Cohen-or. An image is worth one word: Personalizing text-to-image generation using textual inversion. In *The Eleventh International Conference on Learning Representations*, 2022.
30. Nataniel Ruiz, Yuanzhen Li, Varun Jampani, Yael Pritch, Michael Rubinstein, and Kfir Aberman. Dreambooth: Fine tuning text-to-image diffusion models for subject-driven generation. In *Proceedings of the IEEE/CVF Conference on Computer Vision and Pattern Recognition*, pages 22500–22510, 2023.
31. Nupur Kumari, Bingliang Zhang, Richard Zhang, Eli Shechtman, and Jun-Yan Zhu. Multi-concept customization of text-to-image diffusion. In *Proceedings of the IEEE/CVF Conference on Computer Vision and Pattern Recognition*, pages 1931–1941, 2023.
32. Tejas Gokhale, Hamid Palangi, Besmira Nushi, Vibhav Vineet, Eric Horvitz, Ece Kamar, Chitta Baral, and Yezhou Yang. Benchmarking spatial relationships in text-to-image generation. *arXiv preprint* arXiv:2212.10015, 2022b.
33. Martin Heusel, Hubert Ramsauer, Thomas Unterthiner, Bernhard Nessler, and Sepp Hochreiter. Gans trained by a two time-scale update rule converge to a local nash equilibrium. In Isabelle Guyon, Ulrike von Luxburg, Samy Bengio, Hanna M. Wallach, Rob Fergus, S. V. N. Vishwanathan, and Roman Garnett, editors, *Advances in Neural Information Processing Systems 30: Annual Conference on Neural Information Processing Systems 2017, December 4–9, 2017, Long Beach, CA, USA*, pages 6626–6637, 2017. URL https://proceedings.neurips.cc/paper/2017/hash/8a1d694707eb0fefe65871369074926d-Abstract.html.
34. Tim Salimans, Ian J. Goodfellow, Wojciech Zaremba, Vicki Cheung, Alec Radford, and Xi Chen. Improved techniques for training gans. In Daniel D. Lee, Masashi Sugiyama, Ulrike von Luxburg, Isabelle Guyon, and Roman Garnett, editors, *Advances in Neural Information Processing Systems 29: Annual Conference on Neural Information Processing Systems 2016, December 5–10, 2016, Barcelona, Spain*, pages 2226–2234, 2016. URL https://proceedings.neurips.cc/paper/2016/hash/8a3363abe792db2d8761d6403605aeb7-Abstract.html.
35. Tao Xu, Pengchuan Zhang, Qiuyuan Huang, Han Zhang, Zhe Gan, Xiaolei Huang, and Xiaodong He. Attngan: Fine-grained text to image generation with attentional generative adversarial networks. In *2018 IEEE Conference on Computer Vision and Pattern Recognition, CVPR 2018, Salt Lake City, UT, USA, June 18–22, 2018*, pages 1316–1324. IEEE Computer Society, 2018. https://doi.org/10.1109/CVPR.2018.00143. URL http://openaccess.thecvf.com/content_cvpr_2018/html/Xu_AttnGAN_Fine-Grained_Text_CVPR_2018_paper.html.
36. Jack Hessel, Ari Holtzman, Maxwell Forbes, Ronan Le Bras, and Yejin Choi. CLIPScore: A reference-free evaluation metric for image captioning. In *Proceedings of the 2021 Conference on Empirical Methods in Natural Language Processing*, pages 7514–7528, Online and Punta Cana, Dominican Republic, 2021. Association for Computational Linguistics. https://doi.org/10.18653/v1/2021.emnlp-main.595. URL https://aclanthology.org/2021.emnlp-main.595.
37. Maitreya Patel, Tejas Gokhale, Chitta Baral, and Yezhou Yang. Conceptbed: Evaluating concept learning abilities of text-to-image diffusion models. *arXiv preprint* arXiv:2306.04695, 2023.
38. Jaemin Cho, Yushi Hu, Roopal Garg, Peter Anderson, Ranjay Krishna, Jason Baldridge, Mohit Bansal, Jordi Pont-Tuset, and Su Wang. Davidsonian scene graph: Improving reliability in fine-grained evaluation for text-image generation. *arXiv*, 2023.
39. Uriel Singer, Adam Polyak, Thomas Hayes, Xi Yin, Jie An, Songyang Zhang, Qiyuan Hu, Harry Yang, Oron Ashual, Oran Gafni, et al. Make-a-video: Text-to-video generation without text-video data. In *The Eleventh International Conference on Learning Representations*, 2022.

40. Dongchao Yang, Jianwei Yu, Helin Wang, Wen Wang, Chao Weng, Yuexian Zou, and Dong Yu. Diffsound: Discrete diffusion model for text-to-sound generation. *IEEE/ACM Transactions on Audio, Speech, and Language Processing*, 2023.
41. Felix Kreuk, Gabriel Synnaeve, Adam Polyak, Uriel Singer, Alexandre Défossez, Jade Copet, Devi Parikh, Yaniv Taigman, and Yossi Adi. Audiogen: Textually guided audio generation. In *The Eleventh International Conference on Learning Representations*, 2022.
42. Ben Poole, Ajay Jain, Jonathan T. Barron, and Ben Mildenhall. Dreamfusion: Text-to-3d using 2d diffusion. In *The Eleventh International Conference on Learning Representations*, 2023. URL https://openreview.net/forum?id=FjNys5c7VyY.
43. Chen-Hsuan Lin, Jun Gao, Luming Tang, Towaki Takikawa, Xiaohui Zeng, Xun Huang, Karsten Kreis, Sanja Fidler, Ming-Yu Liu, and Tsung-Yi Lin. Magic3d: High-resolution text-to-3d content creation. In *IEEE Conference on Computer Vision and Pattern Recognition (CVPR)*, 2023.
44. Abhinav Gupta, Praveen Srinivasan, Jianbo Shi, and Larry S. Davis. Understanding videos, constructing plots learning a visually grounded storyline model from annotated videos. In *2009 IEEE Computer Society Conference on Computer Vision and Pattern Recognition (CVPR 2009), 20-25 June 2009, Miami, Florida, USA*, pages 2012–2019. IEEE Computer Society, 2009. URL https://doi.org/10.1109/CVPR.2009.5206492.
45. Ali Farhadi, Mohsen Hejrati, Mohammad Amin Sadeghi, Peter Young, Cyrus Rashtchian, Julia Hockenmaier, and David Forsyth. Every picture tells a story: Generating sentences from images. In *Computer Vision–ECCV 2010: 11th European Conference on Computer Vision, Heraklion, Crete, Greece, September 5-11, 2010, Proceedings, Part IV 11*, pages 15–29. Springer, 2010.
46. Yezhou Yang, Ching Teo, Hal Daumé III, and Yiannis Aloimonos. Corpus-guided sentence generation of natural images. In *Proceedings of the 2011 Conference on Empirical Methods in Natural Language Processing*, pages 444–454, Edinburgh, Scotland, UK., 2011. Association for Computational Linguistics. URL https://aclanthology.org/D11-1041.
47. Tsung-Yi Lin, Michael Maire, Serge Belongie, James Hays, Pietro Perona, Deva Ramanan, Piotr Dollár, and C Lawrence Zitnick. Microsoft coco: Common objects in context. In *European conference on computer vision*, pages 740–755. Springer, 2014.
48. Peter Anderson, Xiaodong He, Chris Buehler, Damien Teney, Mark Johnson, Stephen Gould, and Lei Zhang. Bottom-up and top-down attention for image captioning and visual question answering. In *2018 IEEE Conference on Computer Vision and Pattern Recognition, CVPR 2018, Salt Lake City, UT, USA, June 18–22, 2018*, pages 6077–6086. IEEE Computer Society, 2018. https://doi.org/10.1109/CVPR.2018.00636. URL http://openaccess.thecvf.com/content_cvpr_2018/html/Anderson_Bottom-Up_and_Top-Down_CVPR_2018_paper.html.
49. Xinlei Chen, Hao Fang, Tsung-Yi Lin, Ramakrishna Vedantam, Saurabh Gupta, Piotr Dollár, and C Lawrence Zitnick. Microsoft coco captions: Data collection and evaluation server. *arXiv preprint*, abs/1504.00325, 2015. URL https://arxiv.org/abs/1504.00325.
50. Jae Sung Park, Chandra Bhagavatula, Roozbeh Mottaghi, Ali Farhadi, and Yejin Choi. Visualcomet: Reasoning about the dynamic context of a still image. In *In Proceedings of the European Conference on Computer Vision (ECCV)*, 2020.
51. Zhiyuan Fang, Tejas Gokhale, Pratyay Banerjee, Chitta Baral, and Yezhou Yang. Video2Commonsense: Generating commonsense descriptions to enrich video captioning. In *Proceedings of the 2020 Conference on Empirical Methods in Natural Language Processing (EMNLP)*, pages 840–860, Online, 2020b. Association for Computational Linguistics. https://doi.org/10.18653/v1/2020.emnlp-main.61. URL https://aclanthology.org/2020.emnlp-main.61.
52. Eric Horvitz. On the horizon: Interactive and compositional deepfakes. In *Proceedings of the 2022 International Conference on Multimodal Interaction*, pages 653–661, 2022.
53. Mehdi Mirza and Simon Osindero. Conditional generative adversarial nets. *arXiv preprint* arXiv:1411.1784, 2014.

54. Zhenliang He, Wangmeng Zuo, Meina Kan, Shiguang Shan, and Xilin Chen. Attgan: Facial attribute editing by only changing what you want. *IEEE Transactions on Image Processing*, 28 (11): 5464–5478, 2019.
55. Taesung Park, Ming-Yu Liu, Ting-Chun Wang, and Jun-Yan Zhu. Semantic image synthesis with spatially-adaptive normalization. In *Proceedings of the IEEE/CVF Conference on Computer Vision and Pattern Recognition (CVPR)*, June 2019.
56. Zekun Hao, Arun Mallya, Serge Belongie, and Ming-Yu Liu. Gancraft: Unsupervised 3d neural rendering of minecraft worlds. In *Proceedings of the IEEE/CVF International Conference on Computer Vision*, pages 14072–14082, 2021.
57. Wengling Chen and James Hays. Sketchygan: Towards diverse and realistic sketch to image synthesis. In *2018 IEEE Conference on Computer Vision and Pattern Recognition, CVPR 2018, Salt Lake City, UT, USA, June 18–22, 2018*, pages 9416–9425. IEEE Computer Society, 2018. https://doi.org/10.1109/CVPR.2018.00981. URL http://openaccess.thecvf.com/content_cvpr_2018/html/Chen_SketchyGAN_Towards_Diverse_CVPR_2018_paper.html.
58. Han Zhang, Tao Xu, and Hongsheng Li. Stackgan: Text to photo-realistic image synthesis with stacked generative adversarial networks. In *IEEE International Conference on Computer Vision, ICCV 2017, Venice, Italy, October 22–29, 2017*, pages 5908–5916. IEEE Computer Society, 2017. URL https://doi.org/10.1109/ICCV.2017.629.
59. Prafulla Dhariwal and Alexander Nichol. Diffusion models beat gans on image synthesis. *Advances in neural information processing systems*, 34: 8780–8794, 2021.
60. Jonathan Ho and Tim Salimans. Classifier-free diffusion guidance. In *NeurIPS 2021 Workshop on Deep Generative Models and Downstream Applications*, 2021.
61. Diederik P Kingma and Max Welling. Auto-encoding variational {Bayes}. *Int. Conf. on Learning Representations*, 2013.
62. Tsu-Jui Fu, Xin Wang, Scott Grafton, Miguel Eckstein, and William Yang Wang. Sscr: Iterative language-based image editing via self-supervised counterfactual reasoning. In *Proceedings of the 2020 Conference on Empirical Methods in Natural Language Processing (EMNLP)*, pages 4413–4422, 2020.
63. Colin Conwell and Tomer Ullman. Testing relational understanding in text-guided image generation. *ArXiv preprint*, abs/2208.00005, 2022. URL arXiv:2208.00005.
64. Gary Marcus, Ernest Davis, and Scott Aaronson. A very preliminary analysis of dall-e 2. *ArXiv preprint*, abs/2204.13807, 2022. URL arXiv:2204.13807.
65. Evelina Leivada, Elliot Murphy, and Gary Marcus. Dall-e 2 fails to reliably capture common syntactic processes. *ArXiv preprint*, abs/2210.12889, 2022. URL arXiv:2210.12889.
66. Seunghoon Hong, Dingdong Yang, Jongwook Choi, and Honglak Lee. Inferring semantic layout for hierarchical text-to-image synthesis. In *2018 IEEE Conference on Computer Vision and Pattern Recognition, CVPR 2018, Salt Lake City, UT, USA, June 18–22, 2018*, pages 7986–7994. IEEE Computer Society, 2018. https://doi.org/10.1109/CVPR.2018.00833. URL http://openaccess.thecvf.com/content_cvpr_2018/html/Hong_Inferring_Semantic_Layout_CVPR_2018_paper.html.
67. Tobias Hinz, Stefan Heinrich, and Stefan Wermter. Semantic object accuracy for generative text-to-image synthesis. *IEEE transactions on pattern analysis and machine intelligence*, 2020.
68. Jaemin Cho, Abhay Zala, and Mohit Bansal. Dall-eval: Probing the reasoning skills and social biases of text-to-image generative transformers. *ArXiv preprint*, abs/2202.04053, 2022. URL arXiv:2202.04053.
69. Zineng Tang, Ziyi Yang, Chenguang Zhu, Michael Zeng, and Mohit Bansal. Any-to-any generation via composable diffusion. *arXiv preprint* arXiv:2305.11846, 2023.
70. Benjamin Z Yao, Xiong Yang, Liang Lin, Mun Wai Lee, and Song-Chun Zhu. I2t: Image parsing to text description. *Proceedings of the IEEE*, 98(8): 1485–1508, 2010.

71. Andrej Karpathy and Fei-Fei Li. Deep visual-semantic alignments for generating image descriptions. In *IEEE Conference on Computer Vision and Pattern Recognition, CVPR 2015, Boston, MA, USA, June 7–12, 2015*, pages 3128–3137. IEEE Computer Society, 2015. https://doi.org/10.1109/CVPR.2015.7298932.
72. Junnan Li, Dongxu Li, Caiming Xiong, and Steven C. H. Hoi. BLIP: bootstrapping language-image pre-training for unified vision-language understanding and generation. In Kamalika Chaudhuri, Stefanie Jegelka, Le Song, Csaba Szepesvári, Gang Niu, and Sivan Sabato, editors, *International Conference on Machine Learning, ICML 2022, 17–23 July 2022, Baltimore, Maryland, USA*, volume 162 of *Proceedings of Machine Learning Research*, pages 12888–12900. PMLR, 2022b. URL https://proceedings.mlr.press/v162/li22n.html.
73. Ervine Zheng, Qi Yu, Rui Li, Pengcheng Shi, and Anne Haake. Knowledge acquisition for human-in-the-loop image captioning. In *International Conference on Artificial Intelligence and Statistics*, pages 2191–2206. PMLR, 2023.
74. Vishakh Padmakumar and He He. Machine-in-the-loop rewriting for creative image captioning. In *Proceedings of the 2022 Conference of the North American Chapter of the Association for Computational Linguistics: Human Language Technologies*, pages 573–586, 2022.

Retrieval Augmented Modeling

5

At this point in the book, we have discussed fundamental principles of information retrieval, exploring its key elements, and various approaches to achieving effective retrieval, including multimodal retrieval and generative retrieval. Another important application of retrieval is its integration with language models, referred to as retrieval-augmented modeling. In this chapter, we will focus on this paradigm in detail and provide a taxonomy of retrieval-augmented modeling over multiple dimensions. Imagine a language model that not only comprehends the input it is given but also taps into extensive external knowledge sources to provide more informed and accurate responses. This is concisely what retrieval-augmented modeling aims to accomplish. Another advantage of this is that the language models can be much smaller and yet achieve high performance by being able to query a knowledge corpus or search the web to retrieve the relevant information. This essentially means that building larger and larger language models is not the only way to improve performance.

In Sect. 5.1, we will describe different methods in which the retrieved knowledge can be leveraged by a language model to produce accurate responses. This includes augmenting the input to enrich the context, updating the intermediate layers to refine its understanding, and augmenting the output to generate more informed responses. The training of both the retriever model and the language model is a pivotal aspect of retrieval augmented modeling, and in Sect. 5.2, we will explore three strategies for training these models, namely, training them independently, sequentially, or in a joint manner. Retrieval augmented modeling allows language models to be provided with various types of information. In Sect. 5.3, we will delineate these different types of information, such as knowledge, similar examples, and generated context, which can be leveraged to produce informed responses. In the next section, Sect. 5.4, we will focus on the real-world impact of retrieval augmented language models. We will discuss various practical applications such as fact-checking and mitigation of factual 'hallucinations' that can occur with large language models. Finally, in Sect. 5.5, we will shift our focus to leveraging the generation ability of the LLMs for improving retrieval performance.

5.1 Retrieval Augmentation Architecture

In this section, we will learn about various strategies that enable language models to make use of the retrieved information to provide informed responses. We will start with a method in which the **input is augmented with the retrieved information** [1–4]. Then, we will study techniques that **leverage the retrieved information within the intermediate layers of the model**. We will walk through a representative work, RETRO [5] to understand a cross-attention mechanism that integrates the retrieved information in the intermediate layers of the model. Lastly, the retrieved information can also play a role in **augmenting the model's output**. One example of this is the kNN-LMs approach outlined in Khandelwal et al. [6] where a pre-trained language model is extended by linearly interpolating its next word distribution with a k-nearest neighbors (kNN) model. These techniques allow the language model to incorporate the retrieved information into its responses, thereby enhancing its ability to provide contextually relevant and well-informed answers.

5.1.1 Augment the Input

The first strategy to utilize the retrieved information simply augments this information to the given input. This augmentation provides a richer context to the model and helps it in finding a more accurate and informative response.

Consider the task of open-domain question answering in which a question is given as input and the system typically requires external knowledge to find the answer. Here, the retrieved information is used as additional context to answer the given question. Specifically, a system implementing this approach comprises of two modules, a retriever module, and a reader module; thus it is referred to as a **retriever-reader system**. The retriever module is responsible for selecting relevant segments of text (knowledge) from a large database of information in response to a user's query and the reader model leverages that to find the correct answer. Various techniques can be employed for retrieval, and the choice often depends on the specific requirements of the task. Some commonly used retrieval methods include TF-IDF-based sparse retrievers [1], Anserini retriever [2], and dense retrievers [3, 4]. In short, the role of the retriever is to identify and extract pieces of information that are likely to be relevant to the user's query. The retrieved information is then passed on to the reader module which is tasked with comprehending and interpreting this information to predict an accurate answer to the user's question. Different reader models have been used to leverage the retrieved information; for example, Izacard and Grave [4] uses generative model (T5) and Chen et al. [1], Yang et al. [2], Wang et al. [7] use autoencoding BERT models.

One way to augment the input with the retrieved information is to simply concentrate them as pass as input to the model. However, this has a limitation as language models can

5.1 Retrieval Augmentation Architecture

be passed only a limited number of input tokens. This limitation is addressed by the second technique in which each retrieved context is independently encoded and concatenated later at the decoding stage. Next, we will describe these two techniques.

Combined Encoding by Appending to the Input: In this approach, the retrieved information is added as additional context to the input query. By doing this, the model is explicitly given the relevant context to generate a more informed response. However, as previously mentioned, there's a limitation to the amount of information that can be provided, as it cannot exceed the maximum input token length supported by the model (e.g., 512 tokens for BERT or a few thousand tokens for more recent Large Language Models).

Independent Encoding and Concatenation: The retrieved information typically consists of multiple contexts or paragraphs. To address the limitation of processing all the contexts together, in this technique, each context is processed independently. One such method to achieve this feeds each retrieved passage to the document reader independently (along with the input query), and the answer is extracted from each passage separately. The prediction with the highest probability (across all the inferences) is chosen as the final prediction [1].

Another variation of this method considers the similarity scores of passages with the query to weigh the predicted answer from each passage and then selects the one with the highest probability [8] as the final prediction.

These methods process the retrieved passages independently and do not allow interactions between them which could be beneficial in providing accurate responses.

Addressing the limitations of independent encoding, Izacard and Grave [4] proposed a Fusion-in-Decoder (FiD) model that encourages the interaction between the retrieved passages. FiD uses an encoder-decoder transformer model for this purpose. Specifically, each passage is independently encoded along with the query, but the decoder pays attention to the concatenation of the resulting representations of all the retrieved passages. This allows for interaction between the passages and enables the model to perform evidence fusion in the decoder. Figure 5.1 shows the architecture of the FiD model showcasing how it combines the latent information from multiple passages in the decoding stage, improving the model's ability to provide more contextually relevant responses. Building upon the FiD framework, researchers have introduced methods aimed at enhancing the efficiency of inference [9, 10].

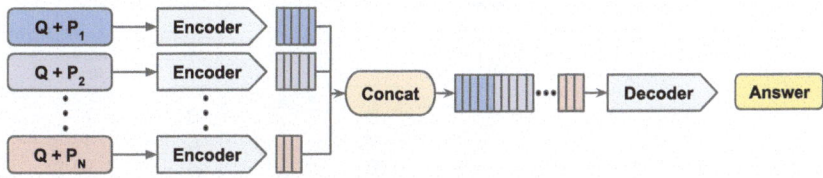

Fig. 5.1 Architecture of the **Fusion-in-Decoder** model [4]

5.1.2 Augment the Intermediate Layers

In this section, we will discuss a technique that utilizes the retrieved information in the intermediate layers of the model. Specifically, we will describe Retrieval-Enhanced Transformer (RETRO) [5] that integrates the retrieved information into the model through a cross-attention mechanism.

At a high level, the RETRO model splits the input sequence into multiple chunks and retrieves text similar to the previous chunk to provide more context and thus improve the predictions in the current chunk.

The original paper used MassiveText dataset for both training and retrieval of data. Specifically, for retrieval, it first constructs a key-value database, where values correspond to raw chunks of text tokens and keys correspond to the frozen BERT embeddings. Each training sequence is then split into chunks, which are augmented with their k-nearest neighbors retrieved from the database. Finally, an encoder-decoder architecture integrates retrieved chunks into the model's predictions.

Consider an n-token-long example $X = (x_1, x_2, ..., x_n)$. It is split into l chunks $(C_1, C_2, ..., C_l)$ of size $m = n/l$. Each chunk C_u is augmented with a set $\text{RET}(C_u)$ of k neighbors from the database. The retrieval is done based on the L_2 distance of BERT embeddings of the chunk and the key in the database. RETRO uses an encoder-decoder architecture to integrate the retrieved data into the model's predictions through a cross-attention mechanism.

Figure 5.2 shows the architecture of the RETRO model. The retrieved tokens are first fed into an encoder transformer, which computes the encoded neighbors set E. Denoting the intermediate activations by H, the transformer decoder then interleaves Retro-blocks Retro(H, E) and standard Transformer blocks LM(H). These blocks are built from three dif-

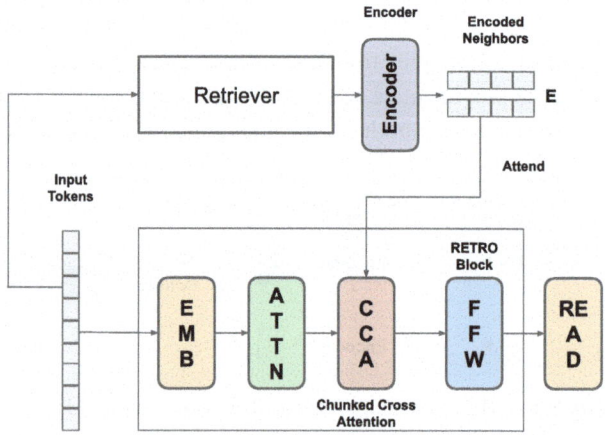

Fig. 5.2 Architecture of the **RETRO model** [5]

ferent residual operators: a fully connected layer, the standard sequence-level self-attention layer, and a chunked cross-attention layer that incorporates information from the retrieval encoder.

5.1.3 Augment the Output

Now, we will turn to the final category of methods for leveraging the retrieved information. In the previous two methods, we augmented the input and the intermediate layers using the retrieved information. In this method, we will augment the output using the retrieved information. A popular method in this category is *k***NN-LMs** [6] that extends a pre-trained language model by linearly interpolating its next word probability distribution with a *k*-nearest neighbors (kNN) model. In summary, the kNN-LMs work by leveraging the nearest neighbors for a given context (identified based on the distance in the embedding space of the pre-trained language model) while generating the next word probability distribution. These nearest neighbors can be sourced from any text dataset, including the data used to train the original language model.

At the core, language models assign probabilities to sequences, i.e., given a context sequence of tokens $c_t = (w_1, ...w_{t-1})$, autoregressive LMs estimate $p(w_t|c_t)$, the distribution over the target token w_t. The *k*NN-LM involves augmenting such a pre-trained language model with a nearest neighbors retrieval mechanism, without any additional training (i.e., the representations learned by the LM remain unchanged). This is done with a single forward pass over a text collection where the resulting context-target pairs are stored in a key-value datastore that is queried during inference. The text collection can potentially also include the original LM training set.

Let's now walk through each component of this method in more detail. Firstly, we will define data store that will be used for retrieval. Then, we will describe the inference task.

Datastore: Let $f(.)$ be the function that maps a context c to a fixed-length vector representation computed by the pre-trained LM. Then, given the ith training example $(c_i, w_i) \in D$, we define the key-value pair (k_i, v_i), where the key k_i is the vector representation of the context $f(c_i)$ and the value v_i is the target word w_i. The datastore (K, V) is thus the set of all key-value pairs constructed from all the training examples in D:

$$(K, V) = \{(f(c_i), w_i)|(c_i, w_i) \in D\}$$

Inference: At the inference time, the language model (LM) is provided with an input context denoted as x, we break down the inference task into the following steps.

Step 1: The LM generates an output distribution that represents the likelihood of the next words, denoted as $p_{LM}(y|x)$, and it also generates a context representation denoted as $f(x)$ that is used for retrieving k-nearest neighbors.

Step 2: To enrich its understanding of the context, the LM uses the context representation $f(x)$ as a query to search the data store. The search is performed to find a set of k-nearest neighbors, which are determined based on a distance function such as the squared L2 distance.

Step 3: After retrieving these nearest neighbors from the data store, the model proceeds to calculate a distribution over these neighbors. This distribution is computed based on a softmax function applied to the negative distances between the context representation $f(x)$ and the representations of the retrieved neighbors. This softmax operation essentially assigns probabilities to the neighbors, where neighbors with smaller distances get assigned higher values.

Step 4: The LM now combines and aggregates the probability mass for each vocabulary token based on its occurrences in the retrieved target neighbors. This means that if a particular word is frequently found in the set of nearest neighbors, it will have a higher probability in the distribution over the next words. Conversely, words that are rarely or never found in the retrieved target neighbors will have a probability of zero in this distribution.

$$p_{\text{kNN}}(y|x) \propto \sum (k_i, v_i) \in N \mathbb{1}_{y=v_i} \exp(-d(k_i, f(x)))$$

Step 5: Finally, the nearest neighbor distribution p_{kNN} is interpolated with the model distribution p_{LM} using a tuned parameter λ to produce the final kNN-LM distribution:

$$p(y|x) = \lambda\, p_{\text{kNN}}(y|x) + (1-\lambda)\, p_{\text{LM}}(y|x)$$

This approach offers the following advantages:

- **It enables the explicit memorization of rare patterns** rather than relying on implicit learning within the model's parameters. This means that the kNN-LM can handle infrequent or unusual language constructs more effectively.
- **It also improves the performance** when the same training data is used for learning both the prefix representation by the language model and the kNN model. This finding suggests that the task of language prediction is more complex than previously recognized, and supplementing it with information from similar contexts can lead to better results.
- **The kNN-LM approach is versatile and supports domain adaptation**. By changing the source of the nearest neighbor data, i.e., the text collection from which k nearest neighbors are retrieved, the model can easily adapt to different domains or specific text sources, offering a valuable tool for tailoring language generation to different contexts or purposes.

5.2 Training of Retrieval Augmented LLM

Equipped with the understanding of different ways in which the retriever can be used with a language model, in this section, we will look at training these modules. Training can be categorized into three classes: **independent training** (in which the retriever and reader are trained independently), **sequential training** (in which one module is fixed and the other is trained upon that fixed module), and **joint training** (in which the retriever and the reader are trained together).

5.2.1 Independent Training

In independent training, the retriever and the reader models are trained independently of each other. These independently trained models allow the system to substitute one or both of these models with other models possessing equivalent functionality without requiring any re-training of the system. For example, a system using the TF-IDF model as a retriever and Fusion-in-Decoder as a reader can replace the TF-IDF retriever with a DPR retriever without requiring re-training of the system.

We have already discussed independent training of retriever models like DPR [11] in Sect. 3.3. Here, we will discuss the independent training of a Reader model. Particularly, we will focus on the Fusion-in-Decoder model [4] which is the state-of-the-art model for open-domain Question Answering.

The **FiD (Fusion-in-Decoder) model** is built upon the T5 model, which is a type of sequence-to-sequence model pretrained on unsupervised data. The FiD model takes as input the question, as well as the retrieved passages, and generates the answer. To achieve this, each retrieved passage, along with its title, is combined with the question, and these combinations are processed separately by the model's encoder. Special tokens, namely "question:", "title:", and "context:", are added before the respective parts of each passage, making it clear which part of the input is the question, which is the title, and which is the passage text.

Once these passages have been encoded individually, the decoder component of the model performs attention over the combined representations of all the retrieved passages. This allows the model to fuse the evidence from various passages and generate a coherent answer. This fusion process takes place exclusively in the decoder, which is why the model is referred to as Fusion-in-Decoder (FiD).

One significant advantage of processing the passages independently in the encoder is that it enables the model to efficiently handle a large number of supporting passages. This is because the encoder only needs to perform self-attention over one context (passage) at a time, resulting in linear growth in computation time as the number of passages increases, rather than quadratic growth. However, by performing evidence fusion in the decoder, the model can better aggregate information from multiple passages, enhancing its ability to provide accurate answers. Figure 5.1 shows the architecture of the FiD model.

5.2.2 Sequential Training

In Sequential training, we keep one module fixed and train the other. Here, we will particularly focus on training the retriever model with frozen LLMs. An important application is the training of a retriever to select optimal demonstrations for Large Language Models (LLMs). Since LLMs have remarkable few-shot learning capability, the quality of these demonstrations plays a crucial role in their overall performance. Luo et al. [12] introduce a method for training a demonstration retriever using insights from LLMs. This system is named Dr. ICL. In the following section, we will detail the process of training the demonstration retriever using the Dr. ICL approach (Fig. 5.3).

Demonstration-Retrieved In-Context Learning: Dr. ICL

In In-context learning (ICL), given a task T an input text x_q, an LLM is used to predict the answer y_q conditioned on a set of *demonstrations* of the task, $Demo = \{d_1, d_2, \ldots, d_n\}$, where $d_i = (x_i, y_i)$ is a pair of input and ground truth answer. Usually, d_i s a text string structured according to a specific template(e.g., "question: x_i backslashn answer: y_i") and then provided to the LM.

Various approaches exist for selecting a set of demonstrations. For example, one might choose a fixed set $Demo$, either randomly or through manual selection, and apply this set uniformly to all queries related to task T. Another option is to employ a retriever to identify demonstrations that are specifically tailored to each query from the training set D_{train}:

$$Demo_{x_q} = Retriever(x_q, D_{train}, n), \qquad (5.1)$$

where $Demo_{x_q}$ are the top-n demonstrations that the retriever considers most suitable for the input x_q. In Dr. ICL, they first consider two off-the-shelf retrievers, BM25 and GTR.

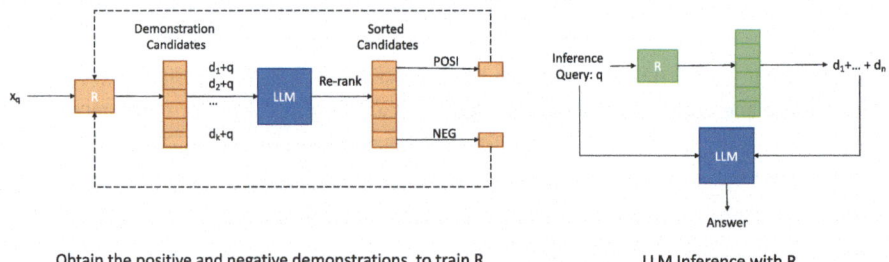

Fig. 5.3 Pipeline: the training and inference process for a demonstration retriever, marked as 'R' for a neural retriever. The left part of the figure illustrates the data acquisition process for training a demonstration retriever: an existing retriever processes an input query x_q and fetches the top-k(for example, 100) demonstration candidates from a training corpus. Subsequently, a Large Language Model (LLM) evaluates the compatibility of each retrieved demonstration with the input query x_q and the ground truth y_q, assigning scores accordingly. The right side of the figure depicts the inference process using the trained demonstration retriever for in-context learning

5.2 Training of Retrieval Augmented LLM

BM25 [13], is a bag-of-words model that determines relevance scores based on term frequency, inverse document frequency, and document length normalization. This model is efficient and effective, making it a practical choice for large-scale, real-world applications. However, BM25 primarily depends on keyword matching and does not fully grasp contextual meanings, which can lead to inaccuracies. In contrast, GTR, introduced by [14], is a dual-encoder neural retriever based on T5 and trained using the MS-MARCO dataset [15]. GTR is superior in understanding semantics and context and can be easily adapted to various downstream tasks and specific fields. Nonetheless, it is less efficient in terms of memory and computation and lacks interpretability. While both BM25 and GTR offer valuable features, they are not specifically trained for demonstration retrieval in the context of Large Language Models (LLMs). Therefore, to enhance retrieval effectiveness, it is recommended to develop a training method for a retriever that is specially tailored to this particular task.

Demonstration Retriever Training The goal of demonstration retrieval is to identify demonstrations that are most relevant and representative for each given input query. Ideally, these demonstrations should encompass two key aspects: (a) specific knowledge pertinent to the query that aids in its resolution, and (b) an understanding of the task's nature and the general approach for solving it.

The goal is to train the retriever model to identify examples that yield the most accurate predictions. The proposed method involves mining a set of demonstrations for each input query x_q in the training dataset. This process begins with a question-answer pair $(x_q, y_q) \in D_{train}$ from the training set D. An off-the-shelf retriever is employed to find a candidate set of demonstrations D for x_q, ensuring x_q is not included in D. The next step involves evaluating each demonstration $d \in D$ for its effectiveness in the target task. This is done by using the LM probability $p_{LM}(y_q \mid d, x_q)$ of the correct answer y_q as the performance score for each demonstration. Finally, the top-n demonstrations are retained as positive examples, while the bottom-n are kept as hard negative examples, effectively creating a balanced set of demonstrations for training purposes.

Training Procedure The retriever is a dual encoder, initialized with GTR pretrained weights. The retriever defines the score of any query-document pair (q, d) as $s(q, d) = v_q^\top v_d$, where v_q and v_d are the embeddings of q and d. It undergoes fine-tuning on the training data using contrastive loss, which incorporates both in-batch negatives and hard negatives:

$$\mathcal{L}_{con} = -\log \frac{e^{s(q,d^+)}}{e^{s(q,d^+)} + \sum_j e^{s(q,d_j^-)}}, \tag{5.2}$$

where d^+ and d_j^- are the positive and negative demonstrations. Negative demonstrations are composed of positive demonstrations from other input queries within the same batch, along with one hard negative demonstration selected at random.

5.2.3 Joint Training

In joint training, as the name suggests, we train both the retriever and reader together. To describe this process, we will look at a representative method called Retrieval Augmented Generation [16].

Retrieval Augmented Generation (RAG): RAG models involve integrating two memory components for generation, parametric and non-parametric memory. The parametric memory refers to a pre-trained sequence-to-sequence model, while the non-parametric memory is a dense vector index containing knowledge from a corpus, which can be accessed using a pre-trained neural retriever. RAG integrates these elements within a probabilistic model that is trained in an end-to-end manner. First, a retriever like DPR supplies latent documents based on the input, and subsequently, a seq2seq model such as BART factors in these latent documents along with the input to produce the output. RAG employs a top-K approximation to marginalize the latent documents, either per output or per token. Furthermore, RAG can be fine-tuned for any seq2seq task, enabling simultaneous learning of both the generator and the retriever components. Figure 5.4 illustrates the RAG architecture.

Let the given input sequence be x, RAG models use the x to retrieve text documents z which are used as additional context for generating the target sequence y. RAG consists of two components:

1. A **retriever** $p_\eta(z|x)$ with parameters η that return (top-K truncated) distributions over text passages from the corpus for the input query x and
2. A **generator** $p_\theta(y_i|x, z, y_{1:i-1})$ with parameters θ that generates a token based on a context of the previously generated $i - 1$ tokens $y_{1:i-1}$, the given input x and a retrieved passage z.

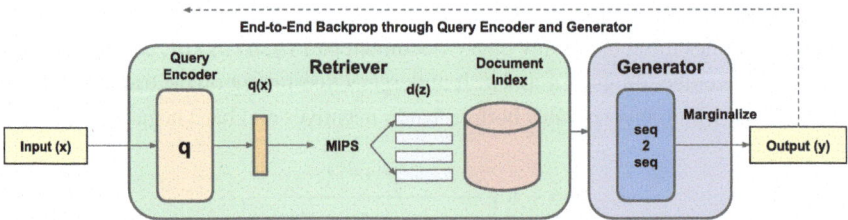

Fig. 5.4 Illustration of **Retrieval Augmented Generation** approach [16]. It combines a pre-trained retriever, comprised of a Query Encoder and a Document Index, with a pre-trained seq2seq model, known as the Generator, and then undergoes end-to-end fine-tuning. For a given query x, it uses MIPS to find the top-K relevant documents z_i. For prediction y, it treats z as a latent variable and marginalizes over seq2seq predictions given different documents

5.2 Training of Retrieval Augmented LLM

To train the retriever and generator in an end-to-end manner, the retrieved document z is treated as a latent variable. RAG models can marginalize the latent documents in two different ways to produce a distribution over the generated text: RAG-Sequence and RAG-Token.

RAG-Sequence: In RAG-Sequence, the model uses the same document to predict each target token (the complete output sequence), i.e., it treats the retrieved document as a single latent variable that is marginalized to get the seq2seq probability $p(y|x)$ via a top-K approximation. Specifically, the retriever finds the top-K documents and the generator produces the output sequence probability for each document, which is then marginalized:

$$p_{\text{RAG-Sequence}}(y|x) = \sum_{z \in \text{top-}k(p(.|x))} p_\eta(z|x) \prod_i^N p_\theta(y_i|x, z, y_{1:i-1})$$

RAG-Token: In the RAG-Token approach, can predict each target token based on a different document. This allows the generator to choose content from several documents when producing an answer. Here, the retriever retrieves the top-K documents and the generator produces a distribution for the next output token for each document, before marginalizing and repeating this process with the following output token,

$$p_{\text{RAG-Token}}(y|x) = \prod_i^N \sum_{z \in \text{top-}k(p(.|x))} p_\eta(z|x) p_\theta(y_i|x, z, y_{1:i-1})$$

The retriever $p_\eta(z|x)$ is based on DPR which consists of a query encoder and a document encoder. The query encoder is finetuned end2end with RAG while keeping the document encoder fixed.

$$p_\eta(z|x) \propto \exp\left(d(z)^T q(x)\right) \qquad d(z) = \text{BERT}_d(z), \quad q(x) = \text{BERT}_q(x)$$

The generator component $p_\theta(y_i|x, z, y_{1:i-1})$ could be modeled using any encoder-decoder like T5 or BART model.

The retriever and generator components are jointly trained without any direct supervision on what document should be retrieved. Given a fine-tuning dataset of (input-output) pairs (x_j, y_j), the negative marginal log-likelihood of each target, $\sum_j -\log p(y_j|x_j)$ is minimized. Updating the document encoder BERT_d during training is costly as it requires the document index to be periodically updated. To avoid this, it can be kept fixed (along with the index), and only the query encoder BERT_q and the generator can be fine-tuned.

5.3 Types of Retrieved Information

In this section, we will discuss different types of information that can be retrieved to output a relevant response to the input query.

Information from a Knowledge Corpus Relevant to the Input Query: This method involves retrieving information from a knowledge corpus that is pertinent to the query at hand. Essentially, it entails accessing a structured or unstructured database of information that may include facts, data, or text documents. This is one of the most common and widely used information utilized for responding to a given query. It is particularly valuable when the input query requires factual or reference-based answers, as it relies on pre-existing knowledge that has been compiled and is available in a structured or searchable form such as the Web. Examples of such knowledge corpus include Wikipedia, PubMed, Textbooks, etc. This type of information is useful in a variety of tasks such as open-domain question answering and fact-checking.

Training Examples Similar to the Input Query: This approach involves retrieving examples similar to the given query. Essentially, when presented with a user query, the model can attempt to respond by leveraging similar examples it has encountered during training. This method is particularly more useful when the test queries pertain to more generalized or commonly occurring topics, as it relies on patterns and associations derived from the training data. This type of information is particularly useful in few-shot in-context learning settings where examples similar to the test input can be retrieved and provided in the context of the LLM.

Generated Context Relevant to the Input Query: Unlike the first method which retrieves relevant information from a knowledge base, this method focuses on generating the relevant context. This generation can be done either simply from the parametric knowledge of a pre-trained language model or from a specially trained context generator model. The language model has been pre-trained on a diverse range of text, enabling it to understand language and context across multiple domains. It can employ its built-in knowledge to provide information or context related to the query, even if the information is not explicitly present in the training data. This approach is valuable when responding to queries that may not have examples similar to them in the training dataset but can benefit from the generated context relevant to them. We will discuss this generative retriever topic in more detail in Sect. 5.5.2.

5.4 Applications of Retrieval Augmented Language Models

Information retrieval plays a crucial role in numerous natural language processing tasks, such as, in open-domain question answering, models need to find the relevant information from a large corpus of text and leverage that to answer the question. Fact-checking a claim requires finding evidence from various information sources. Dialogue systems also require

5.4 Applications of Retrieval Augmented Language Models

access to task-specific or general knowledge to carry out informed conversations. In this section, we discuss several real-world applications where retrieval has an important role to play.

5.4.1 Open Domain Question Answering

Question Answering tasks requires developing systems that can answer questions posed in natural language. In the Open-Domain Question Answering task (ODQA), questions could be about nearly anything relying on world knowledge [17–19]. In ODQA, the challenge is that the context containing relevant information about the question is not provided with the question. This is in contrast to the standard reading comprehension task [20, 21] in which a passage containing the answer span is provided with the question. Thus, ODQA is more realistic and requires the system to retrieve information relevant to the question. This information could be retrieved from unstructured sources (such as web documents, books, news articles, and Wikipedia), structured sources (such as tables, graphs, and knowledge bases), or from different modalities (such as images and videos). Natural Questions [22], TriviaQA [23], and HotpotQA [24] are a few examples of widely popular datasets for open-domain question answering. Figure 5.5 shows a pipeline of a typical open-domain question-answering system.

5.4.2 Use Retriever to Mitigate Hallucinations of LLMs

Large language models such as GPT-3, LLaMA, PaLM, and others have achieved remarkable success in generating fluent and coherent text. However, these models often tend to 'hallucinate', i.e., generate text that seems syntactically sound, fluent, and natural but is factually incorrect, nonsensical, or unfaithful to the provided source input. This hallucination phenomenon critically hampers their reliability and trustworthiness. There are numerous reasons behind this tendency of language models to hallucinate, such as source-reference

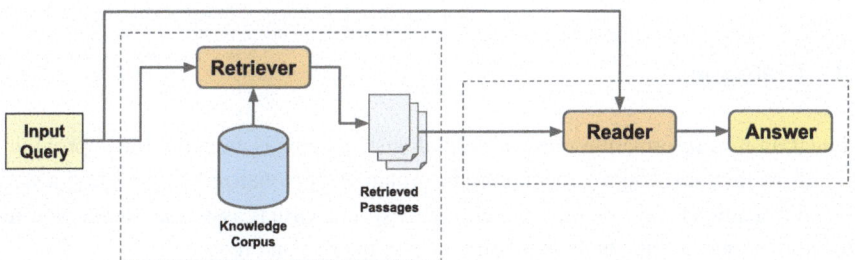

Fig. 5.5 Retriever-Reader pipeline for open-domain question answering

divergence in the pre-training or fine-tuning data, stochasticity or randomness in the decoding techniques, parametric knowledge bias as these models acquire a lot of world knowledge during their pre-training and sometimes they tend to prioritize that knowledge over the contextual information provided to them which can lead to hallucination. Another crucial factor is a discrepancy between training and inference time decoding; specifically, at the inference time the model uses its own generated tokens as the prefix for generating the next token. Deficiency in reasoning capabilities might also contribute to the hallucination problem Banerjee et al. [25], Varshney et al. [26], Luo et al. [27]. So, if it has generated something incorrect then the likelihood of hallucination in the future tokens increases as they will be generated based on incorrect prefixes. Several different kinds of methods have been proposed to address the hallucination problem. One such category of methods uses external knowledge to validate the correctness of the generated content. Specifically, methods such as Active Detection and Mitigation [28], Critic [29], and LLM-Augmenter [30] use knowledge retrieved from external sources such as the web or Wikipedia to validate the generated content. The Active Detection and Mitigation [28] method iteratively generates the sentences and actively intervenes during the generation process when a hallucination is detected; it fixes that hallucination and then proceeds to generate the next sentence. With this active intervention, it aims at preventing the propagation of hallucinations and driving the auto-regressive generation in the right direction. For detecting the hallucination, it first identifies the uncertain concepts using the logit output values of the LLM and then validates the correctness of those concepts using pertinent information retrieved from the web.

5.4.3 Fact Checking

Fact-checking verifies a claim against a collection of evidence. It requires deep knowledge of the claim and reasoning over multiple documents. In the era of abundant information, fact-checking has become an essential tool to combat misinformation and maintain the integrity of information shared on various platforms. Misinformation and related concerns such as disinformation, deceptive news, clickbait, rumors, and information credibility increasingly threaten our society which makes these fact-checking systems crucial.

5.4.4 Dialogue

The integration of information retrieval in dialogue systems is essential for improving the quality of interactions, making them more informative, personalized, and context-aware. It allows conversational AI systems, search engines, and virtual assistants to become more proficient in understanding and responding to user queries and requests.

5.5 Leveraging Generation Ability of LLMs for Better Retrieval

Personalization: By analyzing user preferences and previous interactions, information retrieval can help tailor conversations to the individual. This involves understanding a user's history and delivering content or recommendations that align with their interests and needs.

Knowledge Integration: Dialogue systems can be enhanced by integrating knowledge bases or information sources. Information retrieval techniques can be used to access and extract information from these sources to provide accurate and up-to-date responses during conversations.

Ambiguity Resolution: In complex conversations, where there might be multiple interpretations of a query or response, information retrieval can aid in disambiguation and clarification of the meaning. By accessing relevant data sources, the system can provide contextually appropriate answers.

5.4.5 Slot Filling

The goal of Slot Filling (SF) is to collect information on certain relations (or slots) of entities (e.g., subject entity Albert Einstein and relation educated_at) from large collections of natural language texts. A potential application is a structured Knowledge Base Population. SF requires (1) disambiguation of the input entity and (2) acquiring relational knowledge for that entity. When an information retriever is integrated into the slot-filling process, it assists in the initial step of gathering relevant information from a vast pool of data sources. This retrieved information is subsequently used to populate the designated slots in the input.

5.5 Leveraging Generation Ability of LLMs for Better Retrieval

Now, we will discuss an approach that represents a shift from the traditional retrieval systems that rely on predefined keyword/vector similarity. Specifically, by leveraging the generation ability of LLMs, we can offer a more informed context for information retrieval systems. For instance, in a search engine, when a user enters a query, an LLM can interpret the query in a nuanced way, taking into account the context to provide more accurate results. In this section, we will discuss how we can tap into the generation ability of the language models for better retrieval. First, we will discuss different methods of retrieving knowledge from a language model in Sect. 5.5.1. Then, we will describe different methods to generate query context for better retrieval in Sect. 5.5.2.

5.5.1 Retrieving Knowledge from a Language Model

Large language models pre-trained on unlabeled text achieve remarkable performance when fine-tuned on downstream Natural Language Processing tasks. Furthermore, pre-training enables these models to implicitly store a vast amount of "knowledge" in their parameters. This phenomenon is particularly beneficial for multiple reasons. Firstly, this knowledge is built up by pre-training on unstructured and unlabeled text data which is freely available in huge quantities (in sources such as books and the internet), thus requiring no human supervision for training. Secondly, this parametric knowledge can be retrieved via free-form natural language queries thus requiring no schema engineering, unlike structured knowledge bases. Finally, they support an open set of queries for retrieving the knowledge. In this section, we will first describe LAMA, a benchmark to test the factual and commonsense knowledge of language models. Then, we will detail different methods to elicit knowledge from language models.

5.5.1.1 LAMA Benchmark

Petroni et al. [31] introduced the LAMA (LAnguage Model Analysis) benchmark to test the factual and commonsense knowledge in language models. LAMA includes a variety of sources of factual and commonsense knowledge and consists of a set of facts that are either (subject-relation-object) triples or (question-answer) pairs. Each fact is converted into a cloze statement which is used to query the language model for a missing token. Specifically, a pre-trained language model *knows* a fact (subject, relation, object) such as (Dante, born-in, Florence) if it can successfully predict masked objects in cloze sentences such as "Dante was born in ____" expressing that fact.

LAMA covers four sources of factual and commonsense knowledge, namely, Google-RE, T-REx, ConceptNet, and SQuAD.

Google-RE: The Google-RE corpus[1] contains manually extracted facts from Wikipedia. LAMA considers three relations from Google-RE, namely "place of birth", "date of birth" and "place of death". They define a template for each of these relations, e.g., "[S] was born in [O]" for "place of birth".

T-REx: This knowledge source is a subset of Wikidata triples and is derived from the T-REx dataset [32] which covers a larger and broader set of relations than Google-RE. LAMA includes 41 Wikidata relations and at most 1000 facts per relation. Similar to Google-RE, they manually create templates for each relation.

ConceptNet: ConceptNet [33] is a multilingual knowledge base, initially built on top of Open Mind Common Sense (OMCS) sentences. OMCS represents commonsense relationships between words and/or phrases. LAMA considers facts from the English part of ConceptNet that have single-token objects covering 16 relations. For these ConceptNet triples, LAMA includes a cloze-style question that is created by finding an OMCS sentence that contains both the subject and the object, masking the object in the sentence, and using that sentence as a template for querying language models.

---

[1] https://code.google.com/archive/p/relation-extraction-corpus/.

5.5 Leveraging Generation Ability of LLMs for Better Retrieval

SQuAD: SQuAD [34] is an extractive question-answering dataset. LAMA considers a subset of 305 context-insensitive questions from the SQuAD development set with single-token answers. It includes manually created cloze-style questions from these selected samples, e.g., rewriting "Who developed the theory of relativity?" as "The theory of relativity was developed by _____".

Metric for Evaluation: A model is evaluated based on how highly it ranks the ground truth token against every other word in a fixed candidate vocabulary. This is similar to ranking-based metrics from the knowledge base completion literature. The assumption is that models that rank ground truth tokens high for these cloze statements have more factual knowledge. To this end, LAMA uses rank-based metrics and computes results per relation along with mean values across all relations. To account for multiple valid objects for a subject-relation pair (i.e., for N-M relations), they remove from the candidates when ranking at test time all other valid objects in the training data other than the one we test. It uses the mean precision at k (P@k) i.e. for a given fact, this value is 1 if the object is ranked among the top k results, and 0 otherwise.

5.5.1.2 Methods

Petroni et al. [31] examined the knowledge contained in language models by having the model fill in the blanks of manually created prompts such as "Obama is a _____ by profession". The manually created prompts in Petroni et al. [31] can quite possibly be sub-optimal; for instance, another prompt such as "Obama worked as a _____" may result in more accurately predicting the correct profession as compared to the prompt "Obama is a _____ by profession". To this end, Jiang et al. [35] proposed mining-based and paraphrasing-based methods to automatically generate high-quality and diverse prompts, as well as ensemble methods to combine answers from different prompts. AUTOPROMPT [36] is an automated method to create prompts for a diverse set of tasks, based on a gradient-guided search.

Mining-Based Prompt Generation: This method is inspired by template-based relation extraction methods [37, 40] which are based on the observation that words in the vicinity of the subject x and object y in a large corpus often describe the relation r. Based on this intuition, they first identify all the Wikipedia sentences that contain both subjects and objects of a specific relation r using the assumption of distant supervision, then propose two methods to extract prompts: **Middle-Word Prompts**: Following the observation that words in the middle of the subject and object are often indicative of the relation, we directly use those words as prompts. For example, "Barack Obama was born in Hawaii" is converted into a prompt "x was born in y" by replacing the subject and the object with placeholders.

Paraphrasing-Based Prompt Generation: This method aims to improve the lexical diversity while remaining relatively faithful to the original prompt. Specifically, it paraphrases the original prompt into other semantically similar or identical expressions. For example, if the original prompt is "x shares a border with y", it may be paraphrased into "x has a common border with y" and "x adjoins y". This is similar to query expansion techniques used in information retrieval that reformulate a given query to improve retrieval performance.

5.5.2 Generating Query Context to Improve Retrieval

Query expansion is a widely used technique to improve the effectiveness of information retrieval systems [41]. For tasks like open-domain question answering, augmenting the context of the query facilitates in retrieving more relevant documents that subsequently help the system in accurately answering questions. Techniques like GENREAD [38], GAR [38], and RECITE [39] focus on expanding the query.

Generate-then-Read (GENREAD): GENREAD method prompts a large language model to generate contextual documents for a given question and then reads the generated documents to produce the final answer. Specifically, for generating the contextual documents, they prompt the model with `Generate a background document to answer the given question {question}`.

Generation Augmented Retrieval (GAR): GENREAD uses a large language model to generate the relevant context. Generation Augmented Retrieval (GAR) augments a query for retrieval using a trained 'context' generation model. Specifically, the context generation model is trained with the question as the input and various freely accessible in-domain contexts as the output such as the answer, the sentence containing the answer, and the title of a passage that contains the answer. For a test query, the contexts are generated and appended to the query as the generation-augmented query for retrieval. The expanded query facilitates and makes it easy for the retriever to find the relevant passages.

Recitation-Augmented Generation (RECITE): The RECITE method focuses on a few-shot setting in which a few demonstrations are available. In this method, they leverage the LLM's in-context learning ability and prompt it with paired exemplars of questions and recited evidence. The model can learn from these examples and generate a recitation for a test question. To perform recitation-conditioned few-shot question answering, they append the recited passages at the beginning of the original question-answer exemplars as a single prompt and then generate the final answer. They further use a self-consistency ensembling technique in which top-k sampling is used to independently generate a few recitations, and then greedy decode the answer to the question based on the sampled recitations. Finally, the optimal answer is predicted by taking a majority vote.

5.6 Broader Impact of Retrieval Augmented Modeling

Retrieval Augmented Modeling taps into extensive external knowledge to provide more informed and accurate responses. It has several benefits such as addressing the issue of outdated parametric knowledge, detecting and mitigating hallucinations in generated text, and open-domain dialogue systems. In addition, taking advantage of this technique, the language models can be made much smaller and yet achieve high performance by being able to query a knowledge corpus or search the web to retrieve the relevant information. This means that building larger and larger language models is not the only way to improve the performance. This technique has applications in a variety of real-world tasks including Question Answering, Open-domain Dialogue Generation, and Fact Checking.

Though retrieval-augmented modeling provides several advantages, it has a few downsides as well. Firstly, it increases the inference cost both in terms of computation and latency. Secondly, an important point to note with retrieval-augmented modeling is that its effectiveness is contingent on the quality and veracity of the retrieved knowledge. This is especially critical when retrieving from the web where information is not necessarily correct.

References

1. Danqi Chen, Adam Fisch, Jason Weston, and Antoine Bordes. Reading Wikipedia to answer open-domain questions. In *Proceedings of the 55th Annual Meeting of the Association for Computational Linguistics (Volume 1: Long Papers)*, pages 1870–1879, Vancouver, Canada, 2017b. Association for Computational Linguistics. https://doi.org/10.18653/v1/P17-1171. URL https://aclanthology.org/P17-1171.
2. Wei Yang, Yuqing Xie, Aileen Lin, Xingyu Li, Luchen Tan, Kun Xiong, Ming Li, and Jimmy Lin. End-to-end open-domain question answering with BERTserini. In *Proceedings of the 2019 Conference of the North American Chapter of the Association for Computational Linguistics (Demonstrations)*, pages 72–77, Minneapolis, Minnesota, June 2019. Association for Computational Linguistics. https://doi.org/10.18653/v1/N19-4013. URL https://aclanthology.org/N19-4013.
3. Omar Khattab and Matei Zaharia. Colbert: Efficient and effective passage search via contextualized late interaction over BERT. In Jimmy Huang, Yi Chang, Xueqi Cheng, Jaap Kamps, Vanessa Murdock, Ji-Rong Wen, and Yiqun Liu, editors, *Proceedings of the 43rd International ACM SIGIR conference on research and development in Information Retrieval, SIGIR 2020, Virtual Event, China, July 25–30, 2020*, pages 39–48. ACM, 2020. URL https://doi.org/10.1145/3397271.3401075.
4. Gautier Izacard and Edouard Grave. Leveraging passage retrieval with generative models for open domain question answering. In *Proceedings of the 16th Conference of the European Chapter of the Association for Computational Linguistics: Main Volume*, pages 874–880, Online, 2021. Association for Computational Linguistics. https://doi.org/10.18653/v1/2021.eacl-main.74. URL https://aclanthology.org/2021.eacl-main.74.
5. Sebastian Borgeaud, Arthur Mensch, Jordan Hoffmann, Trevor Cai, Eliza Rutherford, Katie Millican, George van den Driessche, Jean-Baptiste Lespiau, Bogdan Damoc, Aidan Clark, Diego de Las Casas, Aurelia Guy, Jacob Menick, Roman Ring, Tom Hennigan, Saffron Huang, Loren

Maggiore, Chris Jones, Albin Cassirer, Andy Brock, Michela Paganini, Geoffrey Irving, Oriol Vinyals, Simon Osindero, Karen Simonyan, Jack W. Rae, Erich Elsen, and Laurent Sifre. Improving language models by retrieving from trillions of tokens. In Kamalika Chaudhuri, Stefanie Jegelka, Le Song, Csaba Szepesvári, Gang Niu, and Sivan Sabato, editors, *International Conference on Machine Learning, ICML 2022, 17–23 July 2022, Baltimore, Maryland, USA*, volume 162 of *Proceedings of Machine Learning Research*, pages 2206–2240. PMLR, 2022. URL https://proceedings.mlr.press/v162/borgeaud22a.html.
6. Urvashi Khandelwal, Omer Levy, Dan Jurafsky, Luke Zettlemoyer, and Mike Lewis. Generalization through memorization: Nearest neighbor language models. In *8th International Conference on Learning Representations, ICLR 2020, Addis Ababa, Ethiopia, April 26–30, 2020*. OpenReview.net, 2020. URL https://openreview.net/forum?id=HklBjCEKvH.
7. Zhiguo Wang, Patrick Ng, Xiaofei Ma, Ramesh Nallapati, and Bing Xiang. Multi-passage BERT: A globally normalized BERT model for open-domain question answering. In *Proceedings of the 2019 Conference on Empirical Methods in Natural Language Processing and the 9th International Joint Conference on Natural Language Processing (EMNLP-IJCNLP)*, pages 5878–5882, Hong Kong, China, November 2019. Association for Computational Linguistics. https://doi.org/10.18653/v1/D19-1599. URL https://aclanthology.org/D19-1599.
8. Jinhyuk Lee, Seongjun Yun, Hyunjae Kim, Miyoung Ko, and Jaewoo Kang. Ranking paragraphs for improving answer recall in open-domain question answering. In *Proceedings of the 2018 Conference on Empirical Methods in Natural Language Processing*, pages 565–569, Brussels, Belgium, October-November 2018. Association for Computational Linguistics. https://doi.org/10.18653/v1/D18-1053. URL https://aclanthology.org/D18-1053.
9. Neeraj Varshney, Man Luo, and Chitta Baral. Can open-domain qa reader utilize external knowledge efficiently like humans? *arXiv preprint* arXiv:2211.12707, 2022.
10. Michiel de Jong, Yury Zemlyanskiy, Joshua Ainslie, Nicholas FitzGerald, Sumit Sanghai, Fei Sha, and William Cohen. Fido: Fusion-in-decoder optimized for stronger performance and faster inference. *arXiv preprint* arXiv:2212.08153, 2022.
11. Vladimir Karpukhin, Barlas Oguz, Sewon Min, Patrick Lewis, Ledell Wu, Sergey Edunov, Danqi Chen, and Wen-tau Yih. Dense passage retrieval for open-domain question answering. In *Proceedings of the 2020 Conference on Empirical Methods in Natural Language Processing (EMNLP)*, pages 6769–6781, Online, 2020a. Association for Computational Linguistics. https://doi.org/10.18653/v1/2020.emnlp-main.550. URL https://aclanthology.org/2020.emnlp-main.550.
12. Man Luo, Xin Xu, Zhuyun Dai, Panupong Pasupat, Mehran Kazemi, Chitta Baral, Vaiva Imbrasaite, and Vincent Y Zhao. Dr. icl: Demonstration-retrieved in-context learning. *arXiv preprint* arXiv:2305.14128, 2023b.
13. Stephen Robertson, Hugo Zaragoza, et al. The probabilistic relevance framework: Bm25 and beyond. *Foundations and Trends® in Information Retrieval*, 3(4): 333–389, 2009.
14. Jianmo Ni, Chen Qu, Jing Lu, Zhuyun Dai, Gustavo Hernández Ábrego, Ji Ma, Vincent Y Zhao, Yi Luan, Keith B Hall, Ming-Wei Chang, et al. Large dual encoders are generalizable retrievers. *arXiv preprint* arXiv:2112.07899, 2021.
15. Tri Nguyen, Mir Rosenberg, Xia Song, Jianfeng Gao, Saurabh Tiwary, Rangan Majumder, and Li Deng. Ms marco: A human generated machine reading comprehension dataset. In *CoCo NIPS*, 2016.
16. Patrick S. H. Lewis, Ethan Perez, Aleksandra Piktus, Fabio Petroni, Vladimir Karpukhin, Naman Goyal, Heinrich Küttler, Mike Lewis, Wen-tau Yih, Tim Rocktäschel, Sebastian Riedel, and Douwe Kiela. Retrieval-augmented generation for knowledge-intensive NLP tasks. In Hugo Larochelle, Marc'Aurelio Ranzato, Raia Hadsell, Maria-Florina Balcan, and Hsuan-Tien Lin, editors, *Advances in Neural Information Processing Systems*

33: Annual Conference on Neural Information Processing Systems 2020, NeurIPS 2020, December 6–12, 2020, virtual, 2020b. URL https://proceedings.neurips.cc/paper/2020/hash/6b493230205f780e1bc26945df7481e5-Abstract.html.

17. Danqi Chen, Adam Fisch, Jason Weston, and Antoine Bordes. Reading Wikipedia to answer open-domain questions. In *Proceedings of the 55th Annual Meeting of the Association for Computational Linguistics (Volume 1: Long Papers)*, pages 1870–1879, Vancouver, Canada, 2017a. Association for Computational Linguistics. https://doi.org/10.18653/v1/P17-1171. URL https://aclanthology.org/P17-1171.

18. Man Luo, Mihir Parmar, Jayasurya Sevalur Mahendran, Sahit Jain, Samarth Rawal, and Chitta Baral. Sconer: Scoring negative candidates before training neural re-ranker for question answering. In *ICML 2022 Workshop on Knowledge Retrieval and Language Models*, 2022c.

19. Man Luo. Neural retriever-reader for information retrieval and question answering. Technical report, Arizona State University, 2023.

20. Pranav Rajpurkar, Jian Zhang, Konstantin Lopyrev, and Percy Liang. SQuAD: 100,000+ questions for machine comprehension of text. In *Proceedings of the 2016 Conference on Empirical Methods in Natural Language Processing*, pages 2383–2392, Austin, Texas, 2016a. Association for Computational Linguistics. https://doi.org/10.18653/v1/D16-1264. URL https://aclanthology.org/D16-1264.

21. Man Luo, Kazuma Hashimoto, Semih Yavuz, Zhiwei Liu, Chitta Baral, and Yingbo Zhou. Choose your QA model wisely: A systematic study of generative and extractive readers for question answering. In *Proceedings of the 1st Workshop on Semiparametric Methods in NLP: Decoupling Logic from Knowledge*, pages 7–22, Dublin, Ireland and Online, 2022d. Association for Computational Linguistics. https://doi.org/10.18653/v1/2022.spanlp-1.2. URL https://aclanthology.org/2022.spanlp-1.2.

22. Tom Kwiatkowski, Jennimaria Palomaki, Olivia Redfield, Michael Collins, Ankur Parikh, Chris Alberti, Danielle Epstein, Illia Polosukhin, Jacob Devlin, Kenton Lee, Kristina Toutanova, Llion Jones, Matthew Kelcey, Ming-Wei Chang, Andrew M. Dai, Jakob Uszkoreit, Quoc Le, and Slav Petrov. Natural questions: A benchmark for question answering research. *Transactions of the Association for Computational Linguistics*, 7: 452–466, 2019. https://doi.org/10.1162/tacl_a_00276. URL https://aclanthology.org/Q19-1026.

23. Mandar Joshi, Eunsol Choi, Daniel Weld, and Luke Zettlemoyer. TriviaQA: A large scale distantly supervised challenge dataset for reading comprehension. In *Proceedings of the 55th Annual Meeting of the Association for Computational Linguistics (Volume 1: Long Papers)*, pages 1601–1611, Vancouver, Canada, 2017. Association for Computational Linguis. https://doi.org/10.18653/v1/P17-1147. URL https://aclanthology.org/P17-1147.

24. Zhilin Yang, Peng Qi, Saizheng Zhang, Yoshua Bengio, William Cohen, Ruslan Salakhutdinov, and Christopher D. Manning. HotpotQA: A dataset for diverse, explainable multi-hop question answering. In *Proceedings of the 2018 Conference on Empirical Methods in Natural Language Processing*, pages 2369–2380, Brussels, Belgium, 2018. Association for Computational Linguistics. https://doi.org/10.18653/v1/D18-1259. URL https://aclanthology.org/D18-1259.

25. Pratyay Banerjee, Chitta Baral, Man Luo, Arindam Mitra, Kuntal Pal, Tran C Son, and Neeraj Varshney. Can transformers reason about effects of actions? *ArXiv preprint*, abs/2012.09938, 2020. URL arXiv:2012.09938.

26. Neeraj Varshney, Mihir Parmar, Nisarg Patel, Divij Handa, Sayantan Sarkar, Man Luo, and Chitta Baral. Can nlp models correctly reason over contexts that break the common assumptions? *arXiv preprint* arXiv:2305.12096, 2023a.

27. Man Luo, Shrinidhi Kumbhar, Mihir Parmar, Neeraj Varshney, Pratyay Banerjee, Somak Aditya, Chitta Baral, et al. Towards logiglue: A brief survey and a benchmark for analyzing logical reasoning capabilities of language models. *arXiv preprint* arXiv:2310.00836, 2023c.

28. Neeraj Varshney, Wenlin Yao, Hongming Zhang, Jianshu Chen, and Dong Yu. A stitch in time saves nine: Detecting and mitigating hallucinations of llms by validating low-confidence generation. *ArXiv preprint*, abs/2307.03987, 2023b. URL arXiv:2307.03987.
29. Zhibin Gou, Zhihong Shao, Yeyun Gong, Yelong Shen, Yujiu Yang, Nan Duan, and Weizhu Chen. Critic: Large language models can self-correct with tool-interactive critiquing. *arXiv preprint* arXiv:2305.11738, 2023.
30. Baolin Peng, Michel Galley, Pengcheng He, Hao Cheng, Yujia Xie, Yu Hu, Qiuyuan Huang, Lars Liden, Zhou Yu, Weizhu Chen, et al. Check your facts and try again: Improving large language models with external knowledge and automated feedback. *arXiv preprint* arXiv:2302.12813, 2023.
31. Fabio Petroni, Tim Rocktäschel, Sebastian Riedel, Patrick Lewis, Anton Bakhtin, Yuxiang Wu, and Alexander Miller. Language models as knowledge bases? In *Proceedings of the 2019 Conference on Empirical Methods in Natural Language Processing and the 9th International Joint Conference on Natural Language Processing (EMNLP-IJCNLP)*, pages 2463–2473, Hong Kong, China, 2019b. Association for Computational Linguistics. https://doi.org/10.18653/v1/D19-1250. URL https://aclanthology.org/D19-1250.
32. Hady Elsahar, Pavlos Vougiouklis, Arslen Remaci, Christophe Gravier, Jonathon Hare, Frederique Laforest, and Elena Simperl. T-REx: A large scale alignment of natural language with knowledge base triples. In *Proceedings of the Eleventh International Conference on Language Resources and Evaluation (LREC 2018)*, Miyazaki, Japan, 2018. European Language Resources Association (ELRA). URL https://aclanthology.org/L18-1544.
33. Robyn Speer and Catherine Havasi. Representing general relational knowledge in ConceptNet 5. In *Proceedings of the Eighth International Conference on Language Resources and Evaluation (LREC'12)*, pages 3679–3686, Istanbul, Turkey, 2012. European Language Resources Association (ELRA). URL http://www.lrec-conf.org/proceedings/lrec2012/pdf/1072_Paper.pdf.
34. Pranav Rajpurkar, Jian Zhang, Konstantin Lopyrev, and Percy Liang. SQuAD: 100,000+ questions for machine comprehension of text. In *Proceedings of the 2016 Conference on Empirical Methods in Natural Language Processing*, pages 2383–2392, Austin, Texas, 2016b. Association for Computational Linguistics. https://doi.org/10.18653/v1/D16-1264. URL https://aclanthology.org/D16-1264.
35. Zhengbao Jiang, Frank F. Xu, Jun Araki, and Graham Neubig. How can we know what language models know? *Transactions of the Association for Computational Linguistics*, 8: 423–438, 2020. https://doi.org/10.1162/tacl_a_00324. URL https://aclanthology.org/2020.tacl-1.28.
36. Taylor Shin, Yasaman Razeghi, Robert L. Logan IV, Eric Wallace, and Sameer Singh. AutoPrompt: Eliciting Knowledge from Language Models with Automatically Generated Prompts. In *Proceedings of the 2020 Conference on Empirical Methods in Natural Language Processing (EMNLP)*, pages 4222–4235, Online, 2020. Association for Computational Linguis. https://doi.org/10.18653/v1/2020.emnlp-main.346. URL https://aclanthology.org/2020.emnlp-main.346.
37. Deepak Ravichandran and Eduard Hovy. Learning surface text patterns for a question answering system. In *Proceedings of the 40th Annual Meeting of the Association for Computational Linguistics*, pages 41–47, Philadelphia, Pennsylvania, USA, 2002. Association for Computational Linguistics. https://doi.org/10.3115/1073083.1073092. URL https://aclanthology.org/P02-1006.
38. Yuning Mao, Pengcheng He, Xiaodong Liu, Yelong Shen, Jianfeng Gao, Jiawei Han, and Weizhu Chen. Generation-augmented retrieval for open-domain question answering. In *Proceedings of the 59th Annual Meeting of the Association for Computational Linguistics and the 11th International Joint Conference on Natural Language Processing (Volume 1: Long Papers)*, pages 4089–4100, Online, 2021. Association for Computational Linguistics. https://doi.org/10.18653/v1/2021.acl-long.316. URL https://aclanthology.org/2021.acl-long.316.

39. Zhiqing Sun, Xuezhi Wang, Yi Tay, Yiming Yang, and Denny Zhou. Recitation-augmented language models. In *The Eleventh International Conference on Learning Representations*, 2023. URL https://openreview.net/forum?id=-cqvvvb-NkI.
40. Eugene Agichtein and Luis Gravano. Snowball: Extracting relations from large plain-text collections. In *Proceedings of the Fifth ACM Conference on Digital Libraries*, DL '00, page 85–94, New York, NY, USA, 2000. Association for Computing Machinery. ISBN 158113231X. URL https://doi.org/10.1145/336597.336644.
41. Claudio Carpineto and Giovanni Romano. A survey of automatic query expansion in information retrieval. *ACM Comput. Surv.*, 44(1), 2012. ISSN 0360-0300. URL https://doi.org/10.1145/2071389.2071390.

Outlook 6

The word information is derived from the Latin *īnfōrmātiō* which means "formation", "conception", or "education"–the root verb *īnfōrmō* is also used to mean "to form, mold, or give shape to". The etymology of the word *information* reflects the power of this concept to form, mold, and shape opinions, and therefore societies. And indeed, information has been front and center when it comes to constructing narratives for civilizations. In the medieval ages, all over the world, there was an effort by those in power to restrict the common people's access to information and education. Martin Luther, a major figure of the Protestant Reformation in the 16th century, argued for the rights of people to access religious knowledge in their vernaculars–his translation of the Bible in German (instead of Latin from faraway Vatican) had a significant impact on culture in Europe. This effort was wonderfully complemented by the mechanization of the printing press by Gutenberg, which started the "printing revolution".

Of course, the history of printing has a much longer history before Gutenberg–evidence of stencils, seals, stamps, and block printing can be found in Mesopotamia and China as early as the second millennium BCE. Gutenberg's invention substantially improved the efficiency of printing–this invention could be arguably viewed as the precursor to the information revolution of the 21st century. Very quickly, all over Europe, cities had printing shops and by 1500, printing presses throughout Europe had produced more than twenty million books [1]. This technology had a tremendous impact on the society. Scientists could now communicate their ideas widely–this resulted in the establishment of scientific journals. Thus knowledge could be disseminated and democratized easily and people had access to this knowledge through libraries. Ease of access to information has brought significant structural changes to our societies. Today, we have access to information on our phones and computers–through internet search algorithms we can retrieve documents, news articles, images, and records by simply typing in a query. This access is getting even more democratized through speech technology, digital assistants, and AI-powered assistants such as ChatGPT. Information retrieval has been instrumental in the success of the internet as a means for storing, curating, and retrieving information that can be accessed by humans for various uses.

Information Retrieval (IR) is an essential aspect of the internet era. Improvements in IR algorithms directly lead to a better search experience for the end-user. While initial efforts in IR were focused mainly on document retrieval from text queries [2, 3], recent advances in visual representation learning have also enabled image retrieval applications [4–6]. IR also serves as a vital component in many natural language processing tasks such as open-domain question answering [7, 8] and knowledge and commonsense-based question answering [9–12], and more recently in knowledge-based visual question answering [13–16] and image-captioning [17].

Many datasets and IR algorithms have been developed to deal with input queries from a single modality, such as document retrieval from text queries, image retrieval from text queries, text retrieval from video queries, etc. However, in many cases, the query may be multimodal, for instance, an image of a milkshake and a complementary textual description *"restaurants near me"* should return potential matches of nearby restaurants serving milkshakes. Similarly, a patient may be able to present signs such as swelling using photographs and symptoms such as fever using language. Such functionality is desirable in situations where each modality communicates partial, yet vital information about the required output. The ability to use multimodal queries to retrieve relevant, time-sensitive, and life-saving information about diseases, ailments, case studies, treatments, and therapies is highly critical to informed consultation as well as scientific discovery for public health. This book is designed to address this emerging topic of multimodal information retrieval and could be of interest to researchers, practitioners, and students in working on search, retrieval, generation, reasoning, multimedia, and graphics.

While the technology has several exciting applications and a high potential for impact on important problems, there are several challenges associated with the information that lives on the internet. One of the major challenges is the ability to distinguish between true and fake information. Walter Scheirer, in his book *"A History of Fake Things on the Internet"* [18] discusses the history and origins of fake information, including fake news, digital fake images, and many other forms of information deception. Information has always been used as a weapon by malevolent actors–either by spreading, amplifying, and promoting false information, or by restricting access to truth via censorship and isolation. Both of these approaches for the weaponization of information can lead to societal consequences that can manifest in unfortunate societal evils: stereotypes, prejudice, hatred of the "other", alienation, war, and worse. While we develop technologies that make the process of information retrieval faster, easy to use, and equipped with many features and functionalities that improve user experience, we must be wary of potential misuse. Eric Horvitz's Blue Sky Ideas paper on interactive and compositional deep fakes [19] discusses challenges that lay in the path ahead with malicious uses of content generation techniques to impersonate people, to leverage syn-

6 Outlook

thetic content for disinformation, and to construct "synthetic histories". In the absence of mitigations, Horvitz argues, this may potentially lead us unwittingly into a "post-epistemic" world, where "fact cannot be distinguished from fiction". Researchers have also observed that machine learning models run the risk of replicating or even amplifying the societal biases present in training data. Information retrieval models, especially ones trained on large-scale text or image datasets, may also carry various unexpected biases. With more and more modern information retrieval techniques introduced to real-world applications, great concerns have been raised about the potential effect of such biases on societal fairness and equality. Additionally, efficiency remains a critical challenge in multimodal retrieval. Processing data from multiple modalities demands more computational resources than handling unimodal data. Therefore, ensuring the scalability and efficiency of multimodal retrieval systems, particularly when managing large datasets, is a significant technical hurdle.

In recent times, the emergence of large language models like ChatGPT and GPT-4 has initiated a shift in search methodologies from traditional techniques to generative approaches. Traditionally, search systems would sift through databases to find explicit information, requiring users to dig deeper into each relevant document to synthesize answers themselves. In contrast, modern LLMs are now leveraged to directly generate information. This approach potentially saves users' time in acquiring information relevant to the query. The interactive, dialog-based format of LLMs allows users to request clarifications or more detailed information seamlessly. Despite their effectiveness, LLMs do face significant challenges, namely issues of data hallucination and limited recency in their knowledge. Efficiency and time-saving are only beneficial if the information retrieved or generated by these systems is correct.

Hallucination in the context of LLMs refers to the phenomenon where these models generate factually incorrect information while displaying high confidence in their output. A potential solution to mitigate this issue involves the retrieval of explicit, verified information that can be provided as context for the LLMs. This approach not only offers a reference framework to guide the model's responses but also serves as a means to cross-check and validate the accuracy of the information generated by the LLMs. By integrating this retrieval process, the reliability of LLM-generated content can be significantly enhanced, addressing the challenge of hallucination.

Knowledge recency in LLMs refers to the issue where the information stored in these models is limited to what was available up to the point when they were last trained. For instance, the latest iteration of GPT-4 is trained with data available only until April 2023. Consequently, any developments or information emerging post-April 2023 are not inherently known to the model. Given the substantial resources and time required for training LLMs, it is impractical to continually update these models with the latest information. As a result, the training process cannot keep pace with the rapid generation of new information. One effective strategy to address this limitation and ensure LLMs have access to the most current information is through the indexing of new data and its subsequent retrieval. This approach allows LLMs to incorporate and utilize the latest information, thereby enhancing their relevance and accuracy in responding to contemporary queries.

We've explored the significance of retrieval in the era of current LLMs, but it's also crucial to understand the importance of multimodal approaches. To illustrate this, let's consider several specific examples where multimodal data offers advantages over single-modality data.

- *Comprehensive Understanding*: Multimodal retrieval allows LLMs to process and understand information that combines both text and visual elements. This holistic approach is crucial in contexts where understanding is enhanced by the presence of both textual and visual data, such as educational materials, news articles, or social media content.
- *Enhanced User Experience*: In an increasingly digital world, users interact with content that is not just text-based but also includes images, videos, and other multimedia elements. Multimodal retrieval can significantly improve user experience by providing more relevant, accurate, and contextually rich responses.
- *Real-World Applications*: Many real-world applications, such as medical imaging, surveillance, and autonomous vehicles, require the integration of visual data with textual information. Multimodal retrieval plays a crucial role in these applications, enabling more effective and efficient decision-making.
- *Richer Data Insights*: By analyzing both text and images, LLMs can extract deeper insights that might not be apparent from a single mode of data. This ability is particularly valuable in fields like market research, where understanding consumer behavior involves interpreting both written feedback and visual cues.
- *Natural Human Interaction*: Humans naturally use and interpret multiple modes of communication, including speech, text, and visual cues. Multimodal retrieval allows LLMs to mimic this aspect of human interaction, leading to more natural and intuitive user interactions.
- *Overcoming Data Limitations*: Sometimes, the information in one mode may be limited or ambiguous. Multimodal retrieval can overcome these limitations by supplementing the missing or unclear information with data from another mode, leading to more accurate and reliable outcomes.

As we look toward the future, the field of multimodal retrieval stands at a pivotal juncture, poised for transformative growth and innovation. The integration of text and visual data, as seen in the development of advanced language models like ChatGPT and GPT-4, heralds a new era of information processing where depth, context, and nuance are paramount. The potential of these systems to revolutionize user experience, decision-making, and data interpretation in diverse fields such as healthcare and digital interaction is immense. However, this future also brings forth significant challenges as those we discussed before. The evolution of multimodal retrieval will be marked by continuous learning, adaptation, and integration, striving toward models that not only replicate human understanding but also augment it with the vast capabilities of LLMs. The commitment to innovation, ethical considerations, and user-centric approaches will be key drivers shaping the next frontier of multimodal retrieval technologies.

References

1. Lucien Febvre and Henri-Jean Martin. *The coming of the book: the impact of printing 1450–1800*, volume 10. Verso, 1997.
2. Ellen M Voorhees, Donna K Harman, et al. Trec: Experiment and evaluation in information retrieval, vol. 1, 2005.
3. Tri Nguyen, Mir Rosenberg, Xia Song, Jianfeng Gao, Saurabh Tiwary, Rangan Majumder, and Li Deng. Ms marco: A human generated machine reading comprehension dataset. In *CoCo NIPS*, 2016.
4. Bryan A Plummer, Liwei Wang, Chris M Cervantes, Juan C Caicedo, Julia Hockenmaier, and Svetlana Lazebnik. Flickr30k entities: Collecting region-to-phrase correspondences for richer image-to-sentence models. In *Proceedings of the IEEE international conference on computer vision*, pages 2641–2649, 2015.
5. James Philbin, Ondrej Chum, Michael Isard, Josef Sivic, and Andrew Zisserman. Object retrieval with large vocabularies and fast spatial matching. In *2007 IEEE conference on computer vision and pattern recognition*, pages 1–8. IEEE, 2007.
6. James Philbin, Ondrej Chum, Michael Isard, Josef Sivic, and Andrew Zisserman. Lost in quantization: Improving particular object retrieval in large scale image databases. In *2008 IEEE conference on computer vision and pattern recognition*, pages 1–8. IEEE, 2008.
7. Danqi Chen, Adam Fisch, Jason Weston, and Antoine Bordes. Reading Wikipedia to answer open-domain questions. *Proceedings of the 55th Annual Meeting of the Association for Computational Linguistics (Volume 1: Long Papers)*, pages 1870–1879, Vancouver, Canada, 2017b. Association for Computational Linguistics. https://doi.org/10.18653/v1/P17-1171. URL https://aclanthology.org/P17-1171.
8. Vladimir Karpukhin, Barlas Oguz, Sewon Min, Patrick Lewis, Ledell Wu, Sergey Edunov, Danqi Chen, and Wen-tau Yih. Dense passage retrieval for open-domain question answering. In *Proceedings of the 2020 Conference on Empirical Methods in Natural Language Processing (EMNLP)*, pages 6769–6781, Online, November 2020b. Association for Computational Linguistics. https://doi.org/10.18653/v1/2020.emnlp-main.550. URL https://aclanthology.org/2020.emnlp-main.550.
9. Antoine Bordes, Nicolas Usunier, Sumit Chopra, and Jason Weston. Large-scale simple question answering with memory networks. *ArXiv*, arXiv:abs/1506.02075, 2015.
10. Wen-tau Yih, Matthew Richardson, Chris Meek, Ming-Wei Chang, and Jina Suh. The value of semantic parse labeling for knowledge base question answering. In *Proceedings of the 54th Annual Meeting of the Association for Computational Linguistics (Volume 2: Short Papers)*, pages 201–206, Berlin, Germany, August 2016. Association for Computational Linguistics. https://doi.org/10.18653/v1/P16-2033. URL https://aclanthology.org/P16-2033.
11. Kai Sun, Dian Yu, Jianshu Chen, Dong Yu, Yejin Choi, and Claire Cardie. DREAM: A challenge data set and models for dialogue-based reading comprehension. *Transactions of the Association for Computational Linguistics*, 7: 217–231, March 2019. https://doi.org/10.1162/tacl_a_00264. URL https://aclanthology.org/Q19-1014.
12. Alon Talmor, Jonathan Herzig, Nicholas Lourie, and Jonathan Berant. CommonsenseQA: A question answering challenge targeting commonsense knowledge. In *Proceedings of the 2019*

Conference of the North American Chapter of the Association for Computational Linguistics: Human Language Technologies, Volume 1 (Long and Short Papers), pages 4149–4158, Minneapolis, Minnesota, June 2019. Association for Computational Linguisti. https://doi.org/10.18653/v1/N19-1421. URL https://aclanthology.org/N19-1421.

13. Kenneth Marino, Mohammad Rastegari, Ali Farhadi, and Roozbeh Mottaghi. OK-VQA: A visual question answering benchmark requiring external knowledge. In *IEEE Conference on Computer Vision and Pattern Recognition, CVPR 2019, Long Beach, CA, USA, June 16-20, 2019*, pages 3195–3204. Computer Vision Foundation / IEEE, 2019b. https://doi.org/10.1109/CVPR.2019.00331. URL http://openaccess.thecvf.com/content_CVPR_2019/html/Marino_OK-VQA_A_Visual_Question_Answering_Benchmark_Requiring_External_Knowledge_CVPR_2019_paper.html.

14. Aman Jain, Mayank Kothyari, Vishwajeet Kumar, Preethi Jyothi, Ganesh Ramakrishnan, and Soumen Chakrabarti. Select, substitute, search: A new benchmark for knowledge-augmented visual question answering. *Proceedings of the 44th International ACM SIGIR Conference on Research and Development in Information Retrieval*, 2021.

15. Man Luo, Yankai Zeng, Pratyay Banerjee, and Chitta Baral. Weakly-supervised visual-retriever-reader for knowledge-based question answering. *ArXiv*, arXiv:abs/2109.04014, 2021c.

16. Zhengyuan Yang, Zhe Gan, Jianfeng Wang, Xiaowei Hu, Yumao Lu, Zicheng Liu, and Lijuan Wang. An empirical study of gpt-3 for few-shot knowledge-based vqa. *ArXiv*, arXiv:abs/2109.05014, 2021.

17. Fuxiao Liu, Yinghan Wang, Tianlu Wang, and Vicente Ordonez. Visualnews: Benchmark and challenges in entity-aware image captioning. *arXiv preprint* arXiv:2010.03743, 2020.

18. Walter Scheirer. *A History of Fake Things on the Internet*. Stanford University Press, 2023.

19. Eric Horvitz. On the horizon: Interactive and compositional deepfakes. In *Proceedings of the 2022 International Conference on Multimodal Interaction*, pages 653–661, 2022.

SPRINGER NATURE

GPSR Compliance

The European Union's (EU) General Product Safety Regulation (GPSR) is a set of rules that requires consumer products to be safe and our obligations to ensure this.

If you have any concerns about our products, you can contact us on ProductSafety@springernature.com

In case Publisher is established outside the EU, the EU authorized representative is:

Springer Nature Customer Service Center GmbH
Europaplatz 3
69115 Heidelberg, Germany

The manufacturer's authorised representative in the EU is Springer Nature Customer Service Centre GmbH, Europaplatz 3, 69115 Heidelberg, Germany. If you have any concerns regarding our products, please contact ProductSafety@springernature.com

Printed and bound by CPI Group (UK) Ltd, Croydon, CR0 4YY

26/03/2026

02078991-0006